T0406636

Quality Management in Welded Fabrication

Serhii Fomichov · Yevgenia Chvertko ·
Serhii Minakov · Igor Skachkov · Anna Banin

Quality Management in Welded Fabrication

Technical University of Ukraine "Igor Sikorsky Kyiv Poly technic Institute" as textbook for students and post-graduate students majoring in "Materials Science" and Mechanics

Serhii Fomichov
National Technical University of Ukraine
'Igor Sikorsky Kyiv Polytechnic Institute'
Kyiv, Ukraine

Serhii Minakov
National Technical University of Ukraine
'Igor Sikorsky Kyiv Polytechnic Institute'
Kyiv, Ukraine

Anna Banin
Centauri Business Group Inc.
Bellevue, WA, USA

Yevgenia Chvertko
National Technical University of Ukraine
'Igor Sikorsky Kyiv Polytechnic Institute'
Kyiv, Ukraine

Igor Skachkov
National Technical University of Ukraine
'Igor Sikorsky Kyiv Polytechnic Institute'
Kyiv, Ukraine

ISBN 978-3-031-34799-3 ISBN 978-3-031-34800-6 (eBook)
https://doi.org/10.1007/978-3-031-34800-6

Translation from the Ukrainian language edition: "Управління якістю у зварювальному виробництві (Quality Management in Welded Fabrication)" by Serhii Fomichov et al., © Sergiy Fomichov, Yevgenia Chvertko, Ihor Skachkov, Sergiy Minakov, Anna Banin 2022. Published by «Політехніка» КПІ ім. Ігоря Сікорського (Igor Sikorsky Kyiv Polytechnic Institute, Publ. house "Polytechnica"). All Rights Reserved.

This Springer imprint is published by the registered company Springer Nature Switzerland AG
The registered company address is: Gewerbestrasse 11, 6330 Cham, Switzerland

Foreword

Dear fellow welders!

The welding family unites millions of specialists working in all areas of modern manufacturing on land, underwater, and in outer space.

Today, welding belongs to the category of the most popular manufacturing specialties in Europe, North America, Asia, Latin America, Australia, and Africa.

Welding and the related processes are part of the most complex manufacturing technologies based on the fundamentals of mechanics, electrical engineering, physical chemistry, materials science, applied mathematics, computer science, and robotics. This requires a high level of competence from all categories of those associated with welding: manual workers, engineers, and scientists. Scientific and educational literature on welding is constantly being updated. This textbook series for international welding engineers aims to help in the study of physics fundamentals and welding technologies in accordance with the requirements of international standards and the educational requirements of the International Institute of Welding (IIW). The authors illustrated the text throughout the textbook and tried to make it useful to a wide range of specialists, primarily engineers. We hope that the introductory parts of the textbook sections will also help onsite welders to understand the basics of welding and the related processes.

I wish you, dear fellow welders, creative successes, and business achievements in mastering the complex, modern, and very exciting science of welding and the related processes.

Borys Paton
Director of the E.O. Paton Electric
Welding Institute
Kyiv, Ukraine

Introduction

The textbook *Quality Management in Welded Fabrication* is a pilot project to create a series of textbooks for international welding engineers in accordance with the 'IIW Guideline for International Welding Engineers, Technologists, Specialists and Practitioners. Minimum Requirements for Education, Examination and Qualification.' The aim of the project is to present a slice of modern knowledge in the field of welding science, technology, and equipment. The work was spearheaded by Academician Borys Paton.

The textbook consists of four modules:

- Quality management basics. Quality management system.
- Measurement Control and Recording in welding.
- Imperfections and acceptance criteria.
- Non-destructive testing.

Mastering knowledge in quality management is proposed from the standpoint of the functioning of modern enterprise management systems, the implementation of a process approach, understanding and satisfying the requirements and expectations of interested parties. The basics of risk management are provided. Modules for measurements, imperfections, and non-destructive testing are based on the requirements of international standards. The authors set the task to present the material as illustrated and uncomplicated as possible.

The textbook is based on the experience of professors of the National Technical University of Ukraine 'Igor Sikorsky Kyiv Polytechnic Institute' and the Approved Training Body for International Welding Engineers and Technologists of the International Institute of Welding in the development and implementation of management systems in accordance with the requirements of ISO 9001 [2], ISO 14001, ISO 45001, ISO 11462, and ISO 31000, ISO 10012, in the development and implementation of methods and systems of non-destructive testing and technical diagnostics of welded structures, in the training graduate and postgraduate students and international engineers in the field of welding.

The textbook is also intended for:

- graduate and postgraduate students majoring in the field of welding science and technology, as well as non-destructive testing based on bachelors in mechanics, materials science, or electrics,
- university professors specializing in welding science and technology, and
- welding specialists.

Contents

1 Introduction 1
1.1 Quality Management Basics. Quality Management System 2
1.1.1 Quality. ISO 9001. ISO 3834 2
1.1.2 Definition of Interested Parties and Their Requirements 5
1.1.3 Context of the Organization 6
1.1.4 Quality Management Principles 8
1.1.5 Quality Management System 13
1.1.6 Risk Management 59
References 67

2 Measurement, Control and Recording in Welding 69
2.1 Measurement 69
2.1.1 Measurement Accuracy 70
2.1.2 Unity of Measurement 72
2.1.3 Measurement Classification 73
2.1.4 Parameters of Measured Values 75
2.2 Measuring Instruments 76
2.2.1 Metrological Characteristics of Measuring Instruments 76
2.2.2 Measurement Standard, Reference and Working Measuring Instruments 78
2.2.3 Analog Measurement Instruments 79
2.2.4 Digital Measuring Instrument 80
2.3 Measuring Methods 81
2.3.1 Electrical Parameters 81
2.3.2 Pressure Measurement 88
2.3.3 Gas Consumption 91
2.3.4 Welding Speed 93
2.3.5 Temperature (ISO 13916), Humidity, Wind 96
2.3.6 Measuring Humidity 105

2.3.7 Measuring Wind 106
2.4 Calibration, Verification, and Validation of Welding Equipment 107
2.4.1 Periodical Calibration 108
2.4.2 Confirmation of Metrological Characteristics 108
References 111

3 Imperfections and Acceptance Criteria 113
3.1 Types of Weld Imperfections. ISO 6520-1 113
3.1.1 Identification of Imperfections 113
3.1.2 Cracks 114
3.1.3 Cavities 123
3.1.4 Solid Inclusions 125
3.1.5 Lack of Fusion and Penetration 127
3.1.6 Imperfect Shape 128
3.1.7 Other Imperfections 138
3.2 Significance of Imperfections. Acceptance Criteria. Testing Levels 141
3.3 Engineering Critical Assessment Techniques 144
3.3.1 General 144
3.4 Determination of Critical Defect Size by Stress Concentration Factor 145
3.5 Determination of Critical Length of Cracks and Crack-Like Imperfections 148
Literature References 149

4 Non-destructive Testing 151
4.1 Nondestructive Testing Objects and Selection of Methods 151
4.1.1 Nondestructive Testing Objects 151
4.1.2 Selection of NDT Methods Versus Applications 153
4.2 Organoleptic Methods. Visual Testing 155
4.2.1 Fundamentals of Organoleptic Methods 155
4.2.2 Application Area of Visual Testing 155
4.2.3 VT Technique 156
4.2.4 Advantages and Limitations of VT 159
4.3 Ultrasonic Testing 160
4.3.1 Method Fundamentals 160
4.3.2 Ultrasonic Pulsed Echo Technique (UT-PE) 163
4.3.3 Time-Offlight Diffraction Technique (UT-TOFD) 164
4.3.4 Phased Array Ultrasonic Technique (PAUT) 166
4.3.5 UT Technique 169
4.3.6 Advantages and Limitations of UT 171
4.4 Acoustic Emission Method 172
4.4.1 Method Fundamentals 172
4.4.2 Area of Application of AE 175
4.4.3 Acoustic Emission Technique 178

4.4.4 Advantages and Limitations of AE-Method 178
4.5 Penetrant Testing 179
4.5.1 Method Fundamentals 179
4.5.2 Applicability of PT Method 181
4.5.3 PT Technique 181
4.5.4 Advantages and Limitations 182
4.6 Magnetic Particle Testing 183
4.6.1 Method Fundamentals 183
4.6.2 Applicability of Magnetic Particle Testing 184
4.6.3 MT Technique 184
4.6.4 MT Advantages and Disadvantages 187
4.7 Eddy Current Method 187
4.7.1 Method Fundamentals 187
4.7.2 ET Area of Application 190
4.7.3 Eddy Current Testing Techniques 191
4.7.4 Advantages and Limitations of ET 192
4.8 Magnetic Anisotropy Method 192
4.8.1 Method Fundamentals 192
4.8.2 MA Method Area of Application 195
4.8.3 MA Technique 196
4.8.4 Advantages and Limitations of MA Method 197
4.9 Radiographic Testing 198
4.9.1 Method Fundamentals 198
4.9.2 RT Area of Application 204
4.9.3 X-ray and Gamma-Ray Techniques 204
4.9.4 Advantages and Limitations of RT 208
4.10 NDT Procedures. ISO 17635 208
4.11 Documented NDT Information 211
References 212

About the Authors

Serhii Fomichov, Doctor of Science, Professor is First Deputy Director of the Institute of Materials Science and Welding in the National Technical University of Ukraine 'Igor Sikorsky Kyiv Polytechnic Institute';

Director of IIW/EWF Approved Training Body;

Chairman of the Safeguarding Impartiality Committee of Bureau Veritas Ukraine; and

Board Member of Ukrainian Association for Quality.

Dr. Fomichov has 45 years (since 1978) of experience in non-destructive testing and monitoring of technical condition of welded structures, 27 years (since 1996) of quality management experience, and 15 years (since 2008) of experience in training of welding coordinators (International Welding Engineers, Technologists, Specialists, and Practitioners).

He has overseen:

- the development and implementation of non-destructive testing methods, equipment, and systems;
- the development, implementation, and certification of enterprise management systems in accordance with international standards ISO 9001, ISO 14001, ISO 45001, ISO 22000, ISO 50001, and ISO 11462 as well as conducted training programs at more than 500 companies;
- the audit of enterprise management systems; and
- the research and development of nanotechnology for welding and surfacing of low alloy steels. Most of all, he loves his daughters Alina and Julia. Serhii is also an alpine skiing and mountaineering instructor.

Yevgenia Chvertko, Ph.D. is Deputy Director, Professor of the Institute of Materials Science and Welding in the National Technical University of Ukraine 'Igor Sikorsky Kyiv Polytechnic Institute';

Chief Executive in the Authorized Nominated Body of the International Institute of Welding in Ukraine;

Member of working groups of IIW-EWF IAB (International Authorization Board) Group A 'Education, Training and Qualification';

International Welding Engineer; and
International Welding Inspector (comprehensive level).

Prof. Chvertko has 23 years (since 2000) of experience in welding technology development and 12 years (since 2008) of experience in training of welding coordinators (International Welding Engineers, Technologists, Specialists and Practitioners), inspectors, and workers.

She has overseen:

- application of artificial intelligence methods for predicting quality in welding;
- the development, implementation, and certification of quality management systems in accordance with ISO 9001; and
- the assessment of authorized nominated bodies for personnel education, qualification, and certification and for companies' certification.

Serhii Minakov, Ph.D is Professor of the Institute of Materials Science and Welding in the National Technical University of Ukraine 'Igor Sikorsky Kyiv Polytechnic Institute'.

Prof. Minakov has 39 years (since 1984) of experience in non-destructive testing and monitoring of technical condition of welded structures.

He has overseen the development and implementation of non-destructive testing methods, equipment, and systems.

Igor Skachkov, Ph.D. is Professor of the Institute of Materials Science and Welding in the National Technical University of Ukraine 'Igor Sikorsky Kyiv Polytechnic Institute'.

Prof. Skachkov has 42 years (since 1981) of experience in development of welding equipment, automated control of welding and 16 years (since 2004) of quality management experience.

He has overseen:

- development of automated control systems and power sources in welding;
- application of artificial intelligence methods for predicting quality in welding; and
- the development, implementation, and certification of quality management systems in accordance with ISO 9001, statistical process control in accordance with ISO 11462 as well as measurement management systems.

Anna Banin, Master of Science has 15 years (since 2008) of experience in quality assurance, quality control, and quality management systems development and support.

Anna is Founder and Owner of Centauri Business Group Inc.—a US-based corporation that provides support to management system professionals—where she oversees:

- the development of online training courses based on ISO 9001, ISO 14001, ISO 45001, ISO 50001, ISO 9001, and ISO 26000;
- the development of documented information for quality management systems in accordance with ISO 9001.

Abbreviations

ADC	Analog-to-digital converter (Sect. 2.2.4)
AE	Acoustic emission (Sects. 4.4.1, 4.4.2, 4.4.3 and 4.4.4)
BTR	Brittleness temperature ranges (Sect. 3.1.2)
CAR	Corrective actions (Sect. 1.5.13)
CNR	Contrast-noise ratio (Sect. 4.9.3)
CR	Computer radiography (Sect. 4.9.1)
CT	Computer tomography (Sect. 4.9.1)
DDA	Digital detector array (Sect. 4.9.1)
DDF	Dynamic depth focusing (Sect. 4.3.5)
ESW	ElectroSlag welding (Sect. 3.1.4)
ET	Eddy current testing (Sect. 4.7)
EWF	European Welding Federation
FCAW	Flux-Cored Arc Welding (Sect. 3.1.4)
HAZ	The heat-affected zone
IIW	International Institute of Welding
IP	Imaging plate (Sect. 4.9.1)
IQI	Image quality indicator (Sect. 4.9.3)
ITP	Inspection and Test Plan (Sect. 1.5.10.2)
IWE	International Welding Engineer
IWE	International Welding Specialist
IWT	International Welding Practitioner
IWT	International Welding Technologist (Sect. 4.3.4)
LW	Lateral wave (Sect. 4.3.3)
MA	Magnetic anisotropy (Sect. 4.8)
MAG	Metal active gas [welding] (Sects. 3.1.6 and 3.1.7)
MMA	Manual metal arc [welding] (Sects. 3.1.4 and 3.1.6)
MSP	Management system procedure
MT	Magnetic particle testing (Sects. 4.1.2 and 4.6)
NDT	Non-destructive testing (Sects. 4 and 4.10)
OIML	International Organization of Legal Metrology (Sect. 2.1.2)
OSH	Occupational safety and health

PAUT	Phased array ultrasonic technique (Sects. 4.3.4 and 4.3.5)
PDCA	Plan (planning) → Do (execution, maintenance) → Control (control, testing, verification, audit) → Act (corrective actions, improvement).
PT	Penetrant testing (Sects. 4.1.2 and 4.5)
pWPS	Preliminary welding procedure specification
QMS	Quality management system
RT	Radiographic testing (Sects. 4.1.2 and 4.9)
SAW	Submerged arc welding (Sect. 3.1.4)
SWPS	Standard welding procedure specification
TIG	Tungsten inert gas [welding] (Sects. 1.5.9.4 and 3.1.4)
US	Ultrasonic (Sect. 4.3.1)
UT	Ultrasonic testing (Sects. 4.1.2, 4.3 and 4.3.4)
UT-PE	Ultrasonic pulsed echo technique (Sect. 4.3.2)
UT-TOFD	Time-offlight diffraction technique (Sects. 4.3.3 and 4.3.5)
VT	Visual testing (Sects. 4.1.2, 4.2 and 4.3.1)
WPQR	Welding Procedure Qualification Record
WPS	Welding procedure specification (Sects. 2.1, 2.3.1 and 4.2.3)

Chapter 1
Introduction

The textbook "Quality Management in Welding Fabrication" is a pilot project to create a series of textbooks for international welding engineers in accordance with the "IIW Guideline for International Welding Engineers, Technologists, Specialists and Practitioners. Minimum Requirements for Education, Examination and Qualification." The aim of the project is to present a slice of modern knowledge in the field of welding science, technology, and equipment. The work was spearheaded by Academician Borys Paton.

The textbook consists of four modules:

- Quality management basics. Quality management system. Chap. 1
- Measurement, Control and Recording in Welding. Chap. 2
- Imperfections and acceptance criteria. Chap. 3
- Non-destructive testing. Chap. 4

Mastering knowledge in quality management is proposed from the standpoint of the functioning of modern enterprise management systems, the implementation of a process approach, understanding and satisfying the requirements and expectations of interested parties. The basics of risk management are provided. Modules for measurements, imperfections and non-destructive testing are based on the requirements of international standards. The authors set the task to present the material as illustrated and uncomplicated as possible.

The textbook is based on the experience of professors of the National Technical University of Ukraine "Igor Sikorsky Kyiv Polytechnic Institute" and the Approved Training Body for International Welding Engineers and Technologists of the International Institute of Welding in the development and implementation of management systems in accordance with the requirements of ISO 9001 [2], ISO 14001,

S. Fomichov et al., *Quality Management in Welded Fabrication*,
https://doi.org/10.1007/978-3-031-34800-6_1

ISO 45001, ISO 11462, ISO 31000, ISO 10012, in the development and implementation of methods and systems of non-destructive testing and technical diagnostics of welded structures, in the training graduate and post graduate students and international engineers in the field of welding.

The textbook is also intended for:

- graduate and post graduate students majoring in the field of welding science and technology, as well as non-destructive testing based on bachelors in mechanics, materials science, or electrics.
- university professors specializing in welding science and technology.
- welding specialists.

1.1 Quality Management Basics. Quality Management System

1.1.1 Quality. ISO 9001. ISO 3834

The variety and complexity of the physical and chemical effects associated with welding determines the fact that it is impossible to obtain two absolutely identical welded joints. In this case, what is quality and how to ensure it in relation to welding?

ISO 9000 [1] provides the main terms related to quality.

Quality—the degree to which the set of inherent characteristics of the object meets the requirements.

Moreover, the requirements are determined by the parties interested in the organization's activity and may be:

- established—requirements contained in the documented information of the organization (for example, product requirements contained in the customer contracts),
- mandatory—legislative requirements (for example, product safety requirements),
- perceived—requirements that are the generally accepted practice of the organization.

It is important to understand that the quality of a product or service is not a simple set of characteristics. Ensuring the quality of the welded structure does not boil down to improving welding technologies, methods and means of control, and ensuring that welded joints are defect free. Quality is associated with a set of characteristics through the degree of compliance. The degree of compliance, and therefore the level of quality, is determined by the consumer (and other interested parties). Quality is inextricably linked to the consumer. This communication with the consumer and other interested parties is ensured through quality management.

The degree of compliance, and therefore the level of quality, is determined by the customer (and other interested party). Quality is inextricably linked to the consumer.

Communication with customers and other interested parties is ensured through quality management.

Quality Management—is a coordinated activity to control the organization aimed at achieving results with the greatest degree of compliance with the requirements and expectations of customers and other interested parties.

Any activity is most effective if it is carried out in a cycle of four stages: **P**lan (planning) → **D**o (execution, maintenance) → **C**ontrol (control, testing, verification, audit) → **A**ct (corrective actions, improvement). This cycle is called the PDCA cycle or Deming cycle after the name of the author who first formulated it.

Quality Management in accordance with the PDCA cycle consists of four parts (Fig. 1.1):

(1) Quality Planning begins with the formation of a strategy for the development of welding production at the organization, including new technologies, new equipment, new designs. Ongoing planning is being carried out, including the statement of quality goals at all levels of the organization and plans for their implementation. A study of customer requirements, trends in the development of welding production, the activities of competitors is conducted. A decision is made on the quality requirements level. Requirements for welders, operators, welding coordinators are accepted.
(2) Quality Assurance includes assessment and provision of weldability, development of welding technologies, selection of welding equipment and materials. The competence of personnel is ensured. The coordination of welding work is provided. Controlled welding conditions and traceability are ensured.
(3) Quality Control includes NDT, inspection, personnel certification and validation, control of non-conforming products and process outputs.
(4) Quality Improvement is the adoption by management of methods and means of control, equipment upgrades, redistribution of resources (if necessary) based on analysis of decisions to improve welding technologies. Training and personnel competency increase are ensured.

For quality management in welded fabrication, an understanding of the requirements of two international standards is required—ISO 9001 [2] and ISO 3834 [3].

ISO 9001 [2] is the main international quality management standard that defines:

- principles of quality management,
- general requirements for the quality management system and its processes,
- requirements for initiating the application of the basic methods of quality management, including balancing and satisfying the interests of interested parties, managing risks considering the context of the organization, deploying policies, using the PDCA cycle in process control, and improvement methods.

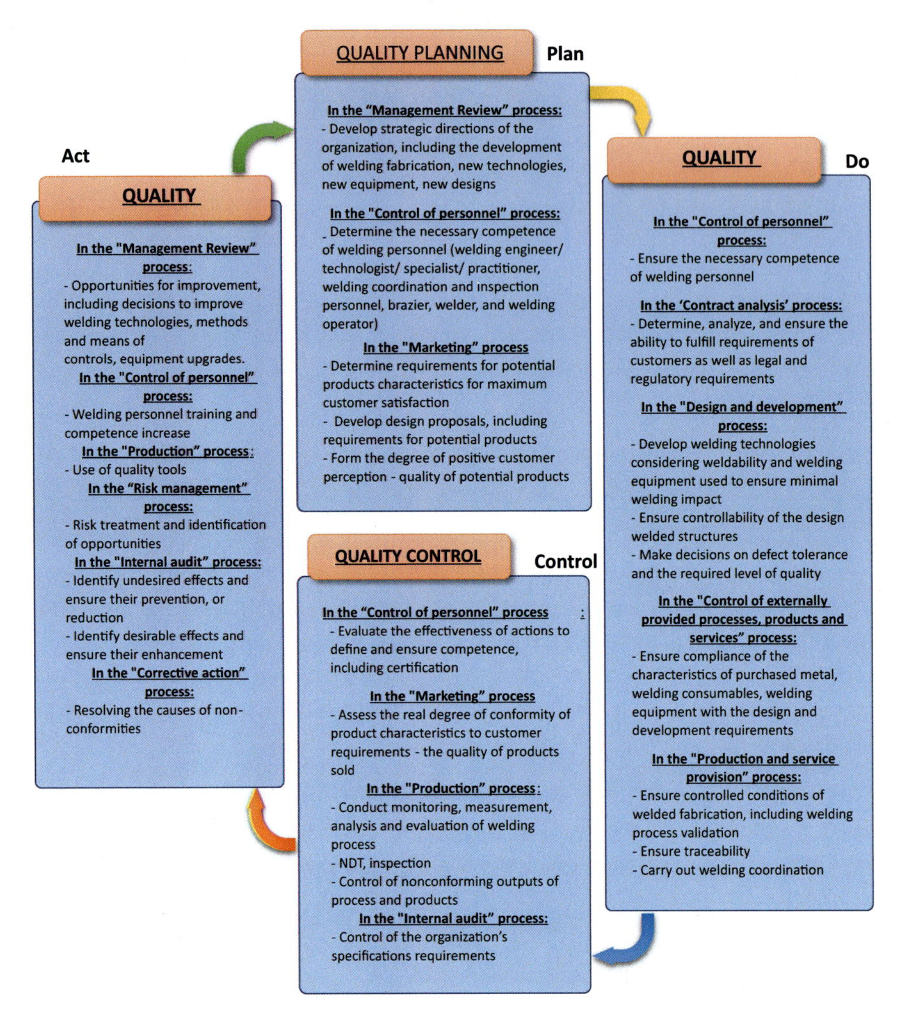

Fig. 1.1 Diagram of Welded Fabrication Quality Management Components in Accordance with PDCA Cycle

ISO 9001 [2] is the universal standard applicable to any organization, regardless of:

- type of production of goods or the provision of services,
- industry affiliation,
- sizes and types of ownership,
- features of national legislation.

ISO 3834 [3] is an auxiliary standard of narrow focus—in relation to welding fabrication. It enables the organization, if necessary, to gradually increase the level of quality, starting from basic requirements, then moving on to standard ones and, finally, rise to meeting the comprehensive quality requirements.

ISO 3834-1 [4] defines three levels of quality requirements:

(a) Comprehensive quality requirements (defined in the second part of ISO 3834-2 [5])—full compliance with the requirements of the ISO 9001 [2] standard is assumed. At the same time, documented information (records) must be kept that indicate that these requirements are met with respect to the main QMS processes (QMS processes are described in detail in Sect. 1.5):

- Control of documented information,
- Management review,
- Control of personnel (including provision of competence),
- Technical maintenance,
- Production,
- Marketing,
- Contract analysis,
- Purchases,
- Internal audit,
- Monitoring, measuring, analysis and evaluation activities.

(b) Typical quality requirements (defined in the third part of ISO 383-3 [6])—full compliance with the requirements of ISO 9001 [2] is assumed. It is not necessary to document the implementation of the requirements.

(c) Basic quality requirements (defined in the fourth part of ISO 3834-4 [7])—established requirements apply only to:

- Welder tests,
- personnel tests for monitoring and testing,
- control of nonconformities and corrective actions.

1.1.2 Definition of Interested Parties and Their Requirements

Requirements that determine the quality of products are formed by interested parties.

Interested party—is a person or other organization that may affect the organization's activities or depend on the organization's activities.

According to the Management system standard requirements, including ISO 9001 [2], the organization should:

- define interested parties,
- define interested parties' requirements relevant to the management system,
- monitor and analyze information about interested parties and their requirements.

Table 1.1 The balance of stakeholders' interests

Interested parties	Interested parties requirements	Organization interests
The owners (shareholders)	Increased capitalization Transparency Sustainable profitability	Investment
Customers	High product quality Low product price On time delivery	Loyalty, high demand
Suppliers and Partners	Mutual benefits and continuity Stability	Mutual benefits and continuity Stability
Personnel	Good working conditions Guarantee of employment Intangible incentives and cash rewards	High labor efficiency
State	Income tax Jobs Compliance with legislative and regulatory requirements	Favorable business environment (legislation, tax system, the legal system)
Society	Environmental protection Jobs Local social development projects	Loyalty

Senior leadership defines who are the interested parties for the organization. In accordance with good practice, the following interested parties are considered:

- Owners (shareholders),
- Customers,
- Suppliers and partners,
- Personnel,
- State,
- Society.

To understand the balance of stakeholders' interests, it is advisable to present the requirements of interested parties in the form of a Table 1.1.

1.1.3 Context of the Organization

Decision making in Quality Management requires an understanding of the context of the organization.

The context of the organization is the environment in which the business is carried out. The context of the organization includes internal and external aspects (Fig. 1.2).

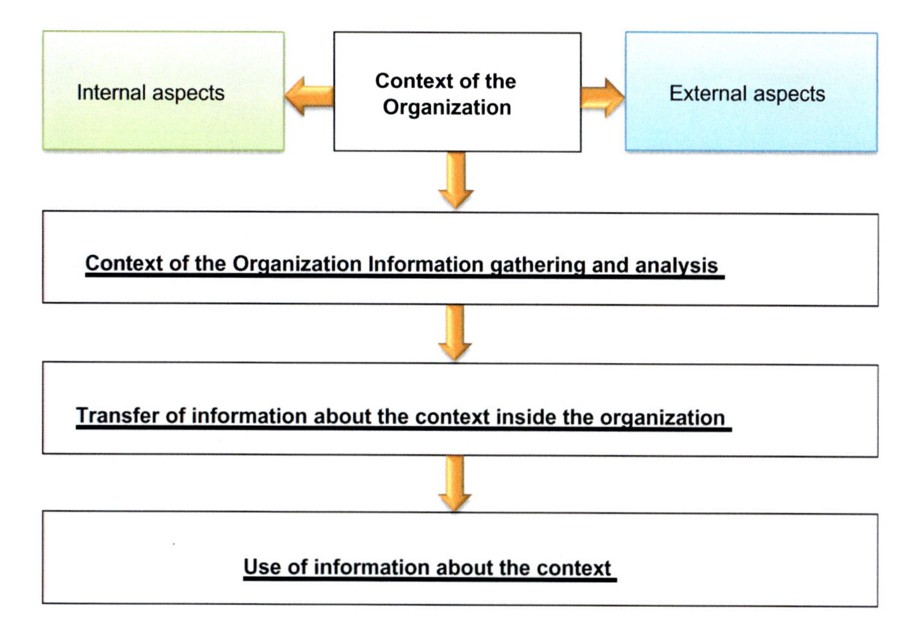

Fig. 1.2 Context of the organization control diagram

Internal aspects include:

- Development strategy,
- Management factors, including organizational structure,
- Performance indicators,
- Technical, technological, and scientific level,
- Production capabilities,
- Resource factors, including infrastructure and production environment,
- State and prospects of development of personnel,
- Cultural and demographic factors,
- Social factors.

External aspects include:

- information from consumers and suppliers,
- competitive factors, including market share of the organization, market leaders' tendencies, market stability,
- macroeconomic factors, such as economic situation, inflation forecast, credit availability,
- external scientific and technological factors,
- political factors, including political stability, public investment, international trade agreements,
- cultural and demographic factors.

Context of the Organization Information gathering and analysis are carried out in the processes: 'Management review', 'Marketing', 'Control of Personnel', 'Design and development of products and (or) services', 'Production and service provision', 'Control of documented Information'.

Transfer of information about the context inside the organization is carried out within the control of organizational knowledge and control of documented information framework.

The use of information about the context is carried out in the QMS processes when planning and making managerial decisions.

1.1.4 Quality Management Principles

Quality management is based on the basic principles that are the foundation of organization's sustainable development:

- Customer focus
- Leadership
- Engagement of people
- Process approach
- Improvement
- Evidence-based decision making
- Relationship management.

Principle 1. Customer focus—making customer satisfaction the highest priority of the organization's activities. The system level is implemented as follows.

The immediate tasks of senior leadership of the organization are:

- ensuring the definition and implementation of customer requirements,
- addressing risks and opportunities associated with customer satisfaction,
- increasing customer satisfaction.

There are three types of activities in the management system that are relevant to customers:

- 'Marketing',
- 'Contract analysis',
- 'Customer communication', including post-delivery activity.

The importance of monitoring and analyses of customer satisfaction information is highlighted by allocation of these requirements into a separate clause of a standard in the general requirements for the management system performance evaluation.

Principle 2. Leadership—is the environment in the organization created by senior management, which provides support for quality management by all members of the organization to achieve the declared objectives. It is implemented via acceptance by senior management of the following obligations:

- Strategic planning, including the development of strategic activities, policies, and objectives of the organization, based on external and internal context, and the integration of standards requirements into business processes,
- Ongoing management review, including provision of resources, performance analysis and promotion of improvements in the management system,
- Demonstrating commitment to the process approach, risk-oriented thinking, effective management and compliance with standards, personnel involvement
- Responsibility for the management system's effectiveness.

Application of Leadership principle ensures:

- considering the needs of all interested parties,
- formation of a clear vision of the organization's future,
- establishing bold objectives and objectives,
- creation and maintenance of common values, impartiality, and the definition of ethical behavior at all levels of the organization,
- creating an atmosphere of trust and work without fear,
- providing the organization's personnel with the required resources, training, and freedom of action within the framework of responsibility,
- inspiration, encouragement, and recognition of employers' contributions.

Principle 3. Engagement of people—is a provision of physical, emotional, and intellectual state that motivates employees to do their job in the best possible way. The engagement of people determines the management system's effectiveness. The task is to make the requirements of the standards a part of the ideology of each employee's activity. This is achieved via:

- Training, ensuring competence,
- Distribution of responsibility and authority,
- Personnel motivation.

Application of Engagement of people principle ensures that employees:

- understand the importance of their contribution and role in the organization,
- identify the factors interfering with their activities,
- accept responsibility for solving problems,
- evaluate their activities in comparison with personal objectives and objectives,
- actively seek improvement of their competence, knowledge, and experience,
- freely transfer their knowledge and experience,
- openly discuss problems and controversial issues.

Principle 4. Process approach—is seeing organization's activities as a set of interrelated and interacting processes.

Process—a set of interrelated activities aimed at converting Inputs to Outputs.

Inputs and Outputs can be material (raw materials, products), information (documented information), undesirable (waste, gas emissions, discharge of liquids). Process—a set of interrelated activities aimed at converting Inputs to Outputs.

Inputs and Outputs can be material (raw materials, products), informational (documented information), undesirable (waste, gas emissions, discharge of liquids).

These processes are the object of management.

Standard management systems, including ISO 9001 [2], have a single process approach implementation program which is carried out in 10 stages.

Stage 1—Define (name) QMS processes.

Stage 2—Define process inputs and outputs.

Stage 3—Determine the sequence and interaction of these processes.

According to the recommendations of ISO/TC 176/SC 2/N 544R [8] *process model is used to visualize internal, external flows and processes interactions. According to BS 6143* [9] *building of process model is done in 4* Steps:

(1) Name the Main Process
(2) Define Outputs (arrows) and Consumer Processes (rectangles with names)
(3) Define Inputs (arrows) Supplier Processes (rectangles with names)
(4) Determine Control Actions (arrows on top of the Main Process with references to standard documentation, based on which the process is performed) and Resources (arrows below the *Main* Process—with specification of the required resources).

An example of a process model is shown in Fig. 1.3.

Step 1—the process is called 'Design and development of products and services'.

Step 2—Process outputs:

- Design and process documentation—to 'Production and Service provision' process.
- Purchase requirements, specifications—to 'Control of externally provided Products and services' process.
- Personnel Competence Requirements—to 'Control of Personnel' process.
- Development report—to 'Management review' process.
- New developments data—to 'Marketing' process.
- Product description and characteristics- to 'Contract analysis' process.

Step 3—Process inputs:

- Design and Development Requirements—from 'Management review' process.
- Market trends analysis, Project requirements—from 'Marketing' process.
- Product Requirements—from 'Contract analysis' process
- Product modernization proposals—from 'Customer communication' process.

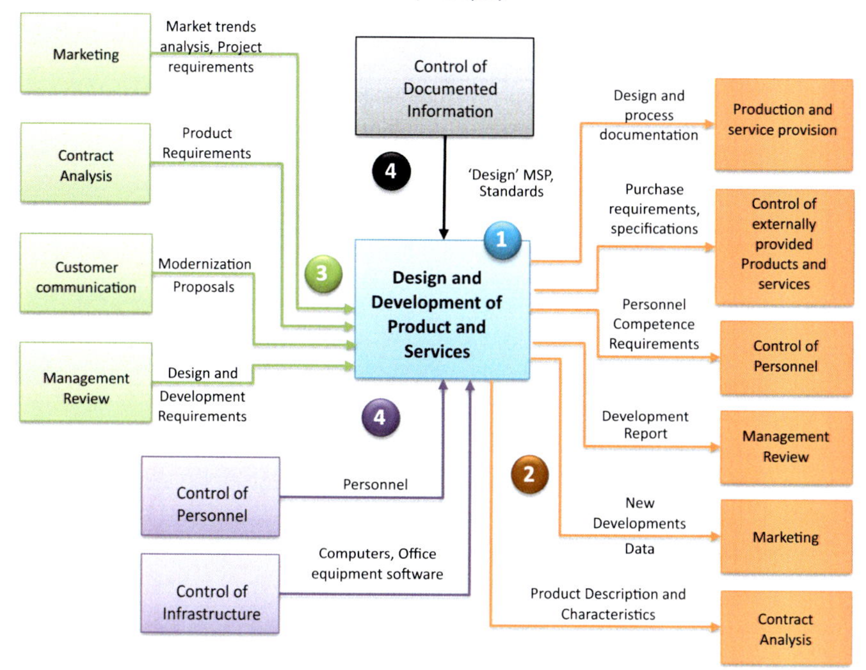

Fig. 1.3 Process model example

Step 4

- Process is managed according to the 'Design and development' MSP and standards from the 'Control of documented information' process.
- Process resources are personnel—from 'Control of personnel' process and computers, office equipment, software from 'Control of infrastructure' process.

Stage 4 of Process approach implementation is to define process criteria and the methods of its measurement.

According to ISO/TC 176/SC 2/N 544R [8] recommendations, the following are the basic two criteria:

- **Effectiveness**—ability to achieve planned results.
- **Efficiency**—ratio of results achieved to the resourced used.

Effectiveness	Efficiency
Ability to achieve planned results	The balance between the results achieved and the resources used
$E = I_a/I_p$	$\mathbf{P = I_a/I_r}$

(continued)

(continued)

Effectiveness	Efficiency
Ia—indicator of achieved result Ip—indicator of planned result	**Ir**—Indicator of resources used

Stage 5—Define resources for process execution.
Stage 6—Distribute responsibility and authority for the processes.
Stage 7—Identify risks and opportunities.
Stage 8—Conduct monitoring and measuring of process criteria.
Stage 9—Improve the process.
Stage 10—Maintain documented information.

Principle 5. Improvement of products, processes, and management system—is the adaptation to the ever-changing requirements of all interested parties. Improvement is a prerequisite for business existence. Business is dynamic. If there is no movement forward, the rollback begins.

System improvement of processes is carried out by performing the following for each process:

- Process criteria activities along the chain: monitoring > measurement > analysis of trends > elimination of causes of negative effects and development of causes of positive effects.
- Addressing risks and opportunities.

Management system improvement is carried out via:

- Breakthrough projects.
- Small steps improvements.

Application of Improvement principle ensures:

- Coordinated approach to improve operations deployment across the organization,
- Employees training in the ways and methods of improvement,
- Turning the improvement of products, processes, and management system into the task of every employee of the organization,
- Developing objectives for improvements guiding and monitoring,
- Recognition and confirmation of improvements.

Principle 6. Evidence-based decision making—based on established facts to reduce the risks of economic and other losses. This is ensured by:

- availability of each management system process criteria,
- application of methodology for determining process criteria based on quality management methodologies, including statistical methods,
- reliability of evidence through the use of appropriate resources for monitoring and measurement.

Evidence-based decision making is implemented:

- when analyzing and evaluating the conformity of products and services, the degree of customer satisfaction, the effectiveness of the management system, the effectiveness of planning, the indicators of external suppliers,
- when conducting internal audits,
- in management review.

Application of Evidence-based decision-making principle ensures:

- Confidence in data and information reliability and accuracy,
- Availability of data for those who need it,
- Data and information analysis using reliable methodologies,
- Making decisions and taking actions based on actual analysis, balanced with experience and intuition.

Principle 7. Relationship management—is the control of the organization's interaction with all interested parties at all levels. Interested parties are individuals and legal entities that create added value for the organization or are somehow interested in the activities of the organization or are under its influence.

This provides the establishment of interested parties relevant to the management system and ensures a balance of interests between each stakeholder and the interests of the organization.

Application of Relationship management principle ensures:

- establishing relationships with external providers and other interested parties based on the balance of short-term achievements of long-term plans,
- combining knowledge, experience, and resources with partners,
- identification and selection of major external providers,
- transparent and open exchange of information, including plans for the future,
- collaboration in development and improvement activities,
- inspiration, encouragement, and recognition by interested parties of improvements and achievements of the organization.

1.1.5 Quality Management System

1.1.5.1 General

Quality Management requires certain processes to be completed. Processes require resources. The processes are performed by personnel whose relationships are determined by the structure. Stable functioning of the processes in accordance with the expectations of interested parties is ensured by the documented information.

Quality Management System (QMS)—the set of interconnected elements (processes, organizational structure, resources, and documented information—Fig. 1.4) necessary for Quality Management, including development of policy, goals and the achievement of these goals based on Quality management principles.

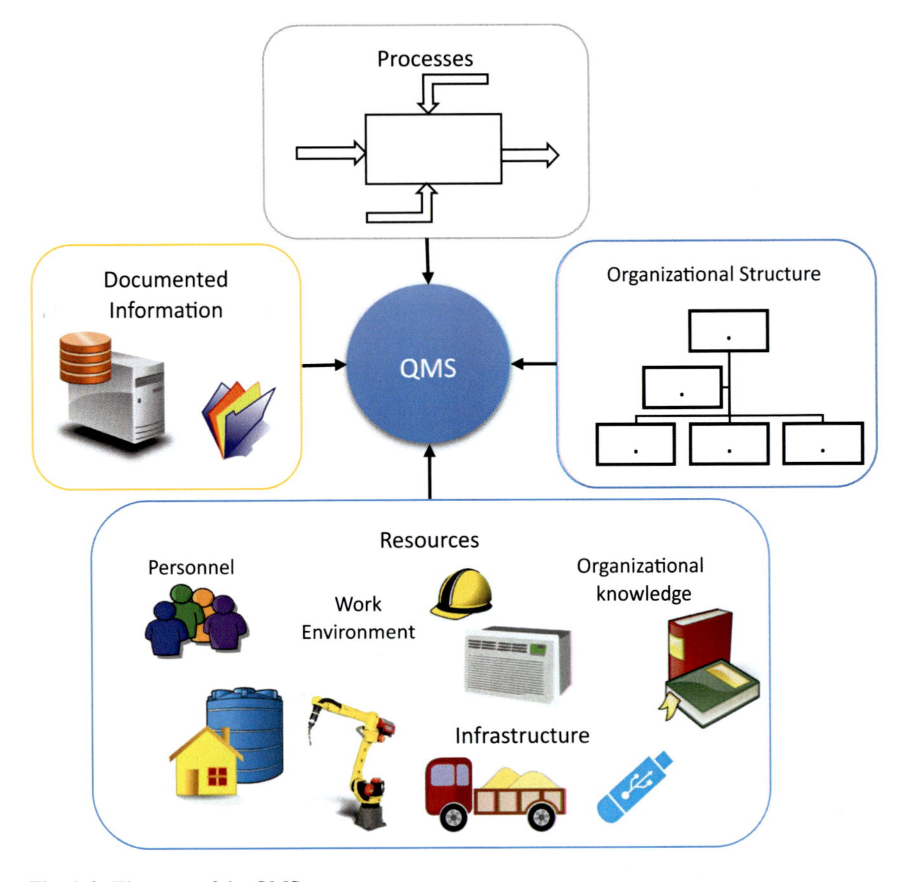

Fig. 1.4 Elements of the QMS

The scope of the QMS are products and processes (the boundaries) that are carried out in accordance with the requirements of standards and the standards (applicability) that the management system corresponds to.

The scope is defined and updated by Senior Leadership based on analysis of (Fig. 1.5):

- Context of the Organization,
- Needs and expectations of interested parties,
- Risks relevant to internal and external aspects.

The scope should be maintained as documented information with reference to:

- Products and services within the QMS,
- Justification of exceptions when the requirement of any clause of the standard cannot be applied.

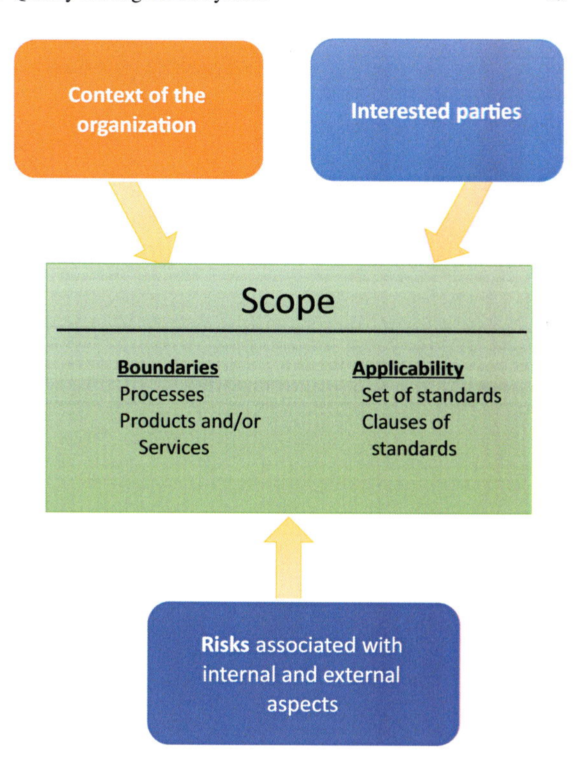

Fig. 1.5 Determining the scope of the QMS diagram

Processes are the core element of the QMS.

Interrelated process models are the business model of the organization (Fig. 1.6). In the example below, the outputs and inputs are simplified.

Let's review the basic processes of a quality management system in the welding industry.

1.1.5.2 Management Review

General

Management Review is a key quality planning process.

Management Review has two levels:

- Strategic planning,
- Ongoing review.

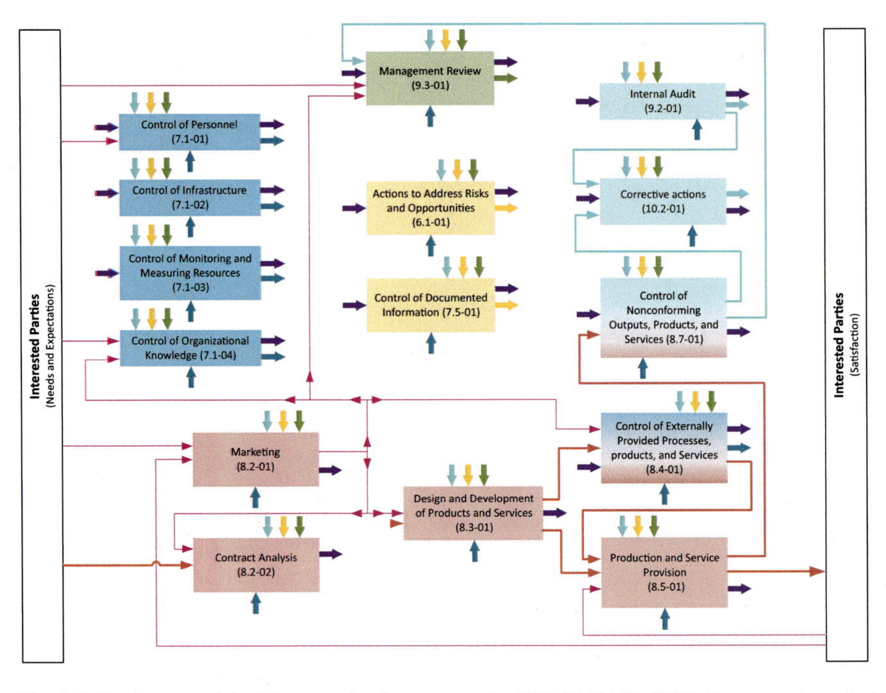

Fig. 1.6 Business model of the organization—example (ISO 9001 [2]: 2015 clause numbers that govern the process requirements are given in parenthesis)

During **Strategic Planning**, based on interested parties' needs and expectations understanding and internal and external context of organization analysis, Top management develops:

- **Mission**—the main common goal of the organization, the meaning of its existence.
- Strategic directions of the organization—a long-term plan of actions that determines the priorities of strategic tasks, resources, and the sequence of steps to achieve a mission, including the development of welding fabrication, new technologies, new equipment, new designs.
- **Quality Policy**—a concise statement of the main intentions and principles of the organization's development in the field of Quality Management.
- **Quality Objectives**—specific, measurable targets for improving operational performance to ensure process conformity and interested parties' satisfaction.

In accordance with good practice, the mission includes concise wording of:

- Purpose (or orientation of the activity) of the organization,
- Market orientation,
- Main competitive advantage,
- List of interested parties.

During **the Ongoing Review**, in accordance with the requirements of ISO 9001 [2], Management Review Agenda includes:

- consideration of the actions' status on the decisions from previous management reviews,
- analysis of changes in external and internal issues that are relevant to the QMS,
- analysis of trends in customer satisfaction and feedback from relevant interested parties,
- analysis of trends to the extent to which Quality Objectives have been met,
- analysis of trends process performance and conformity of products and services,
- analysis of trends nonconformities and corrective actions,
- analysis of trends monitoring and measurement results,
- analysis of audit results,
- analysis of trends in the performance of external providers,
- analysis of the adequacy of resources,
- analysis of the effectiveness of actions taken to address risks and opportunities,
- opportunities for improvement.

Decisions on the results of the current review, in accordance with the requirements of ISO 9001 [2], focused on:

- opportunities for improvement, в including decisions to improve welding technologies, methods and means of control, equipment upgrades,
- any need for changes to the QMS,
- resource needs.

Quality Policy

Based on the Strategic directions of the organization, a set of standards is selected, the requirements of which the organization's management system must meet, and a policy is worded.

In accordance with good practice, the policy has the structure shown in Fig. 1.7 and includes:

- Wording of obligations to implement the management principles, in relation to the characteristics of the organization, including continuous improvement,
- Commitments on risk management as part of decision making,
- Obligations to develop an information security system,
- A list of selected management systems standards, including ISO 9001 [2] and an obligation to implement their requirements.

In accordance with ISO 9001 [2] The Quality Policy must be documented, understandable and accessible to interested parties, including personnel.

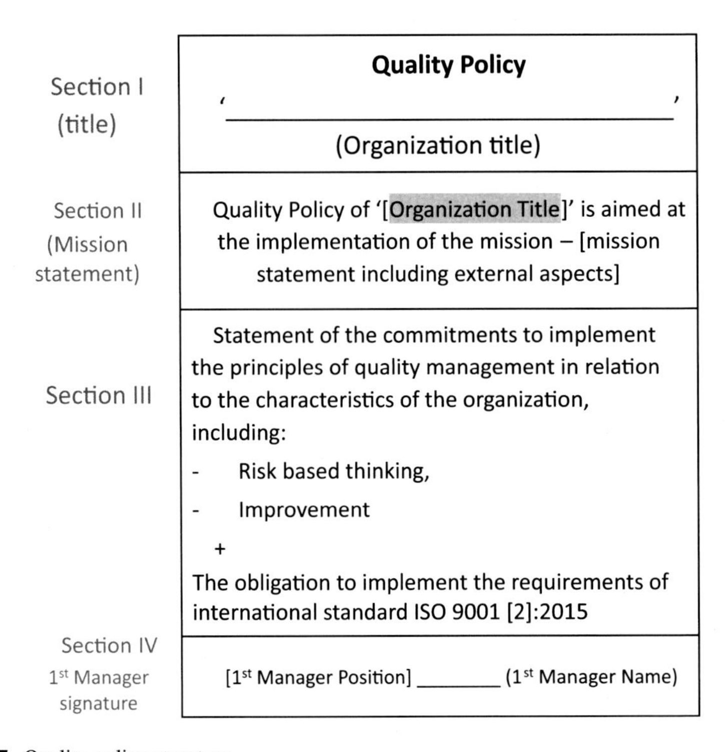

Fig. 1.7 Quality policy structure

Quality Objectives

Based on the Strategic directions of the organization and Quality Policy, Quality Objectives are adopted at various levels of the organization. In accordance with ISO 9001 [2] Quality Objectives should:

- be documented and be measurable,
- relate to ensuring the conformity of products and services and increasing customer satisfaction,
- be monitored,
- be brought to the attention of the staff,
- be updated when necessary.

In accordance with good practice, documenting objectives and planning their achievement is done through the implementation of the organization's Annual development program (Fig. 1.8).

Organization's Annual Development Program

#	Content of activities	Subdivision	Cost If applicable	Responsible	Deadline	Reporting
1	2	3	4	5	6	7
1.	(Quality objectives wording)					
1.1 1.2 …	(Actions to achieve the goals)					

Fig. 1.8 Organization's Annual Development Program Form

The Program includes:

- necessary actions,
- allocation of resources required to deploy the necessary activities,
- distribution of responsibility and authority to achieve each goal,
- a time frame for each goal,
- how the results will be evaluated.

Quality objectives at the subdivision (or process) level include two components:

(1) Subdivision objectives, resulting from the organization's Annual development program.
(2) Objectives aimed at developing and improving activities, based on the resources of the subdivision (or process).

Thus, as a result of the organization's policy deployment, each employee will work towards the interested parties' satisfaction.

1.1.5.3 Control of Personnel

ISO 9001 [2] Requirements for Control of Personnel

The objectives of the 'Control of Personnel' process are:

- determination of the required competence,
- recruitment, hiring, transfer and dismissal of personnel,
- training, including QMS awareness,
- evaluation of effectiveness of measures to ensure competence,
- preservation of documented competence evidence,
- analysis of human resources adequacy, improvement actions.

Main requirements of ISO 9001 [2] standard to personnel management are:

(a) determine the necessary competence of personnel,
(b) ensure the necessary competence of personnel,
(c) evaluate the effectiveness of the actions aimed at determining and ensure the competence sought,
(d) preserve of documented competence evidence.

Competence—combination of appropriate education, training, and experience.
Employees' competence requirements are defined based on:

- the organizational Charter,
- the strategic directions of development,
- QMS and business processes configuration,
- information on the organization's new developments,
- legal requirements,
- requirements of National and International standards used by the company (in case of welding production such standards may be ISO 14731 [10], ISO14732 [11], ISO 9606 series [12–16], ISO 9712 [17], etc.).

Job Descriptions contain sections with the employees' competence requirements.

Education is a component of ensuring personnel competence and is carried out in three target directions.

(a) Promising direction of business, engineering, and technology development,
(b) Continuous professional development of personnel,
(c) QMS training, including:

    - Training programs on management methodologies and new standards,
    - Ongoing employees' training.

Documented competence evidence are:

- certificates of basic education,
- certificates of training, evidence of continuous education, internships, etc.,
- Letters of recommendation.

Activities of Welding Engineer/Technologist/Specialist/Practitioner in the Different Industry Functions in Industry (ISO 14731 [10])

International Welding **E**ngineers, **T**echnologists, **S**pecialists and **P**ractitioners (IIW/EWF-IAB Guideline 252 [18]) are defined as personnel with qualification for welding coordination. General information about their activities and responsibilities is given in the ISO 14731 [10] standard.

According to IIW/EWF-IAB Guideline 252 [18] knowledge, skills and competence levels of mentioned personnel categories are as follows:

IWE—advanced knowledge and critical understanding of welding technology application demonstrating technology mastery, ability to solve high-level complex and unpredictable problems, to manage high complex technical and professional activities or projects related to welding applications, taking responsibility for decision making in unpredictable work or study context and for managing professional development of individuals and groups.

IWT—overall knowledge and understanding of welding technology application demonstrating ability to solve low-level complex problems, to manage in detail welding and related applications or projects as well as professional development of individuals and groups, taking responsibility for decision making in low-level complex work or study context and to define the tasks of welding or related personnel.

IWS—specialized and factual knowledge in the field of welding technology demonstrating ability to develop solutions on common/regular problems, to manage and supervise common or standard welding and related applications, taking responsibility for decision making in common or standard work and in supervising the tasks of welding and related personnel.

IWP—basic knowledge in the field of welding technology demonstrating ability to develop solutions on basic and specific problems, to supervise basic welding and related applications, taking responsibility for decision making in basic work and in supervising the tasks of welding and related personnel.

In correlation with coordination tasks listed in ISO 14731 [10] the previously mentioned competences and skills enable personnel categories to effectively perform the following.

IWE—comprehensive level (eligible to work with all types of welded structures):

- to review Welding construction contract requirements and technical documentation of welded structure.
- to specify, develop, supervise, and manage:
  - subcontracting activities,
  - welding and related personnel competencies and skills,
  - all types of equipment,
  - manufacturing plan and welding procedures, working instructions,
  - basic materials, welding consumables and heat treatment (if applicable),
  - inspection testing plan and corrective actions,
  - identification, traceability, and quality records.

IWT—specific level—tasks listed for IWE limited by structures with low level of complexity.

IWS—basic level—tasks listed for IWE limited by regular and common structures.

IWP (eligible for basic specific works):

- to supervise implementation and monitor:
  - subcontracting activities,
  - welding and related personnel competencies and skills,
  - basic materials, welding consumables and heat treatment (if applicable),
  - inspection testing plan and corrective actions.
- to implement:
  - working instructions,
  - welding procedures;
- to control and perform identification, traceability, and quality records.

Welding Coordination and Inspection Personnel. Qualification tasks and Responsibilities (ISO 14731 [10], ISO 9712 [17])

Welding coordination—coordination of manufacturing operations for all welding and welding-related activities (can be assigned to a person or to a team).

Welding coordination personnel (also referred to as welding coordinators)—a person or group of people performing defined welding coordination tasks. Manufacturer may appoint different personnel for different welding and related tasks at his own discretion, taking into consideration requirements to qualification and/or practical experience (where applicable).

Welding inspection—conformity evaluation of welding variables by observation and judgment accompanied as appropriate by measurement or testing. Welding inspection is a part of welding coordination (*this definition makes welding coordinator responsible for quality assurance in manufacturing of welded structures*).

All welding coordination personnel shall be able to demonstrate:

- competence in the welding-related tasks allocated to them,
- technical knowledge in welding and related technologies relevant.

Further welding coordinators are allocated to one of three specific levels: comprehensive, specific, or basic.

At the **comprehensive level**, welding coordination personnel shall have highly specialized problem-solving skills helping them to define/develop the best technical and economical solutions when applying welding and related technologies for highly complex and unpredictable conditions.

At the **specific level**, welding coordination personnel shall have advanced problem-solving skills. These skills include critical evaluation to select the appropriate technical and economical solutions when applying welding and related technologies, for complex and unpredictable conditions.

At the **basic level**, welding coordination personnel shall have fundamental problem-solving skills. These skills include the ability to identify and develop appropriate solutions, when applying welding and related technologies, for common basic and specific problems.

ISO 14731 [10] gives the following list of essential welding and related tasks (note that in each particular case the concrete list of tasks will change, so the manufacturer may not need welding coordination personnel to perform all of them):

- Review of requirements:
- the product standard to be used, any supplementary requirements.
- the capability of the manufacturer to meet the prescribed requirements.
- Technical review:
- the parent material(s) specification and welded joint properties.
- the joint location with relation to the design requirements.
- quality and acceptance requirements for welds.
- the location, accessibility, and sequence of welds, including accessibility for inspection and nondestructive testing.
- other welding requirements, e.g., batch testing of consumables, ferrite content of weld metal, ageing, hydrogen content, permanent backing, use of peening, surface finish, weld profile;
- the dimensions and details of joint preparation and completed weld.
- Sub-contracting (suitability of each subcontractor for the purpose and their ability to comply with relevant requirements and standards).
- Welding personnel (qualification and compliance with the requirements).
- Equipment:
- the suitability of welding and associated equipment.
- auxiliary equipment supply, identification, and handling.
- personal protective equipment and other safety equipment, directly associated with the applicable manufacturing process.
- equipment maintenance.
- equipment verification and validation.
- Production planning:
- reference to the appropriate procedure specifications for welding and related processes.
- structures in which the welds are to be made.
- environmental conditions (e.g., protection from wind, temperature, and rain).
- the allocation of competent personnel.
- equipment for preheating and post-heat treatment, including temperature indicators.
- the arrangement for any production test.
- Qualification of the welding procedures (method and range of qualification and all variables).
- Welding procedure specifications (range of qualifications shall be considered and conformance with the relevant contract requirements).
- Work instructions.

- Welding consumables:
- compatibility.
- delivery conditions.
- supplementary requirements in the welding consumable purchasing specifications, including the type of welding consumable inspection document.
- the storage and handling of welding consumables.
- batch testing.
- Materials:
- supplementary requirements in the material purchasing specifications, including the type of inspection document for the materials.
- the weldability of the materials to be used.
- the storage and handling of parent material.
- traceability.
- Inspection and testing before welding:
- the suitability and validity of welders' and welding operators' qualification certificates.
- the suitability of the welding procedure specification.
- the identity of the parent material.
- the identity of welding consumables.
- joint preparation (e.g., shape and dimensions).
- fit up, jigging and tacking.
- special requirements in the welding procedure specification (e.g., prevention of distortion).
- the suitability of working conditions for welding, including the environment.
- Inspection and testing during welding:
- essential welding parameters (e.g., welding current, arc voltage and travel speed).
- the preheating/interpass temperature.
- the cleaning and shape of runs and layers of weld metal.
- back gouging.
- the welding sequences.
- the correct use and handling of welding consumables.
- control of distortion.
- intermediate examination (e.g., checking dimensions).
- Inspection and testing after welding:
- the use of visual inspection.
- the use of non-destructive testing.
- the use of destructive testing.
- the form, shape, tolerance, and dimensions of the construction.
- the results and records of post-operations (e.g., post-weld heat treatment, ageing).
- Post-weld heat treatment.
- Non-conformance and corrective actions.
- Calibration and validation of measuring, inspection, and testing equipment (necessary methods and actions).
- Identification and traceability:
- the identification of production plans.

- the identification of routing cards.
- the identification of weld locations in construction.
- the identification of non-destructive testing procedures and personnel.
- the identification of the welding consumable (e.g., designation, trade name, manufacturer of consumables and batch or cast numbers).
- the identification and/or traceability of parent material (e.g., type, cast number).
- the identification of the location of repairs.
- the identification of the location of temporary attachments.
- traceability for fully mechanized and automatic welding units to specific welds.
- traceability of welder and welding operators to specific welds.
- traceability of welding procedure specifications to specific welds.
- Quality records (preparation and maintenance of the necessary records).
- Health and safety and environment (relevant rules and regulations).

Appointment of welding coordination personnel is fully a manufacturer's responsibility, and it is not transferable to other manufacturers, i.e. your customer cannot command your company to appoint a certain person as a responsible coordinator or to change requirements to coordinator's qualification.

Manufacturer's job descriptions for all welding coordination personnel shall include at least their tasks and responsibilities as well as extent of authorization. The manufacturer shall also determine the level of education, qualification and experience required for welding coordination personnel. Competence of welding coordination personnel to follow their assigned tasks is usually checked via assessment, usually including:

- previous experience of welding similar products with the standards used by the manufacturing organization.
- extent of experience fabricating the materials used by the manufacturer.
- previous experience in using the welding supporting standards used by the manufacturing organization (WPQR, WPS, welder and welding operator qualifications, etc.).
- understanding of applicable standards (ISO 3834 series [3–7], ISO 14731 [10] etc.).
- experience of troubleshooting of welding related problems.
- knowledge of relevant essential welding and related tasks.
- theoretical knowledge relevant to the manufacturer.

Non-destructive testing is an essential part of quality assurance in welding manufacturing. However, welding coordination personnel meeting requirements of ISO 14731 [10] standard is not automatically eligible to perform NDT activities.

Qualification and certification of NDT personnel is described in ISO 9712 [17] standard. It is applicable to proficiency in one or more of the following methods: acoustic emission testing, eddy current testing, infrared thermographic testing, leak testing (hydraulic pressure tests excluded), magnetic particle testing, penetrant testing, radiographic testing, strain testing, ultrasonic testing, visual testing (excluding direct unaided visual tests and visual tests carried out during the application of another NDT method). Procedures listed in the standard can be applied to other NDT methods as well.

Similarly, to ISO 14731 [10], ISO 9712 [17] presents three levels of NDT personnel, level 3 being the highest.

Level 3—competence to perform and direct non-destructive testing operations for which the person is certified. Level 3 personnel demonstrate competence to evaluate and interpret results in terms of existing codes, standards, specifications, and procedures as well as sufficient practical knowledge of applicable materials, fabrication and process technology to select NDT methods, establish NDT techniques, to assist in establishing acceptance criteria where none are otherwise available, and a general familiarity with other NDT methods. Within the scope of the competence Level 3 certified specialist may be authorized by the employer to:

- assume full responsibility for a test facility or examination center and staff.
- establish, review for editorial and technical correctness and validate NDT instructions and procedures.
- interpret codes, standards, specifications, and procedures.
- designate the particular test methods, procedures and NDT instructions to be used, carry out and supervise all tasks at all levels.
- provide guidance for personnel at all levels.

Level 2—competence to perform non-destructive testing according to established procedures. Within the scope of the competence defined on the certificate, Level 2 personnel may be authorized by the employer to:

- select the NDT technique for the test method to be used.
- define the limitations of application of the testing method.
- translate NDT codes, standards, specifications, and procedures into NDT instructions adapted to the actual working conditions.
- set up and verify equipment settings.
- perform and supervise tests.
- interpret and evaluate results according to applicable codes, standards, specifications, or procedures.
- prepare NDT instructions.
- carry out and supervise all tasks at or below Level 2.
- provide guidance for personnel at or below Level 2.
- report the results of non-destructive tests.

Level 1—competence to carry out NDT according to NDT instructions and under the supervision of Level 2 or Level 3 personnel. Within the scope of the competence defined on the certificate, Level 1 personnel may be authorized by the employer to:

- set up NDT equipment.
- perform the tests.
- record and classify the results of the tests.
- Report on the results.

Level 1 certified personnel shall not be responsible for the choice of test method or technique to be used, nor for the assessment of test results.

Standard lists competencies for personnel *certified* to a specific level, meaning the document conforming eligibility of inspector to perform certain activities has a pre-defined period of validity which must be prolonged if needed according to procedure defined by the standard.

Qualification and Certification of NDT Personnel (EN ISO 9712 [17])

The candidates shall fulfill the minimum requirements of vision and training prior to the qualification examination and shall fulfill the minimum requirements for industrial experience prior to certification.

Vision requirements apply to all levels. The candidate shall provide documented evidence of satisfactory vision in accordance with requirements. Subsequent to certification, the tests of visual acuity shall be carried out annually. They are to be verified by the employer or the responsible agency.

Training is obligatory for Level 1 and Level 2. There shall be documented evidence, in a form acceptable to the certification body, that training in the method and level for which the certification is sought has been satisfactorily completed.

For Level 3 certification candidates, preparation for qualification may be done in different ways. They may attend training courses, conferences or seminars, study books, periodicals, and other specialized materials. Regardless of the manner of preparation, the Level 3 candidate shall submit documentary evidence of appropriate training. The minimum duration of training depends on the NDT method and is stated in the standard.

Industrial experience may be acquired either prior to or following success in the qualification examination. Documentary evidence of experience shall be confirmed by the employer. If experience is gained after a successful examination, the results of the examination remain valid for up to five years.

Requirements to duration of experience depend on NDT method. However, a reduction is possible, at the discretion of certification bode, taking into account the quality of experience (skills may be assimilated more quickly in an environment where the experience is concentrated), number of NDT processes which candidate is working with (e.g., gaining experience simultaneously in two or more surface NDT methods or same method applied to different sectors). Graduation from technical college or university as well as completion of at least two years of engineering or science study at college or university, may provide justification for a reduction in experience.

Generally, industrial experience provides a reduction of obligatory training up to 50% of its initial duration.

The **qualification examination** covers a given NDT method as applied in one industrial sector, or one or more product sectors. Examinations for levels 1 and 2 include specific (theoretical) and practical parts, both related to the NDT method combined with sector of application. Level 3 must successfully pass basic and main-method examination.

Candidates who successfully completed the activities listed above are eligible for **certification**. Certificate validity period is no longer than 5 years, after which it can be prolonged for one more validity period. After 2 validity periods (10 years) expiration the person should undergo re-certification procedure.

Brazer and Welder Approval/Qualification (ISO 9606 Series)

ISO 9606 series [12–16] consists of the following parts, under the general title "*Qualification test of welders—Fusion welding*":

- Part 1: Steels [12].
- Part 2: Aluminum and aluminum alloys [13].
- Part 3: Copper and copper alloys [14].
- Part 4: Nickel and nickel alloys [15].
- Part 5: Titanium and titanium alloys, zirconium, and zirconium alloys [16].

The principle of this standards series is that a qualification test qualifies the welder not only for the conditions used in the test, but also for all joints which are considered to weld easier on the presumption that the welder has received a particular training and/or has industrial practice within the range of qualifications.

ISO 9606 [12–16] series applies to the following welding processes (designations according to EN ISO 4063 [19]):

- 111—Manual Metal-Arc Welding.
- 114—Self Shielded Flux-Cored Arc Welding.
- 121—Submerged Arc Welding with solid wire electrode.
- 125—Submerged Arc Welding with solid wire electrode (partially mechanized).
- 131—Metal Arc Inert Gas Welding.
- 135—Metal Arc Active Gas Welding.
- 136—Flux-Cored Arc Welding with active gas shield.
- 138—Metal Arc Active Gas Welding.
- 141—Tungsten Inert Gas (arc) Welding (wolfram electrode) with solid -wire or -rod and inert gas.
- 142—Tungsten Inert Gas (arc) Welding without filler material.
- 143—Tungsten Inert Gas (arc) Welding with flux-cored -wire or -rod.
- 145—Tungsten Inert Gas (arc) Welding with solid -wire or -rod with deoxidizing gas (partly).
- 15—Plasma Arc Welding.
- 311—Oxy-Acetylene Gas welding.

Welder's qualification is based on the main parameters, each having a predefined authorization range. If a welder must perform welding outside his/her current qualification, new qualification is required. Main parameters are:

- welding process/processes.
- product type (plate or pipe).
- type of weld (fillet or butt).
- welding technique (where applicable).
- base material or filler material group (see clarification below).
- filler material type (solid wire, flux-cored wire, electrode, rod, etc.).
- dimensions (material thickness and outer pipe diameter).
- welding position (ISO 15614-1 [20]).
- Weld details (single- or multi-run, left- or right-handed welding, single- or double-side joint, backing plate, temporary backing, flux backing).

Qualifications according to Part 1 of the standard (steels) are identified by filler materials group. These groups are: fine-grained and non-alloyed steels, high-strength steels, heat resistant steels Cr < 3,75, heat-resistant steels 3,75% $\leq$ Cr $\leq$ 12%, stainless and ovenproof steels, nickel and nickel-based alloys. Other parts of the standard use identification by base material group.

Group and sub-group of base material according to EN ISO/TR 15608 [21], which were used for qualification, need to be stated in the certificate even if Part 1 is used.

Every qualification as a rule gives permission to weld with one process certain product type (plates or pipes) with certain joint types (fillet or butt) using certain welding techniques. Any change requires a new qualification, however there are several exceptions listed in the standard.

The welder's qualification begins from the date of welding of the test piece(s), provided that the required testing has been carried out and the test results obtained were acceptable.

The Welder's certificate is issued for the period of no more than 3 years (depending on the part applied) and must be prolonged each 6 months by the responsible welding coordinator. After validity period expires welders should undergo re-certification procedure unless other ways of certification period prolongation are eligible.

Brazing and Welding Operator Qualification (ISO 14732 [11])

Welding operators and weld setters for mechanized and automatic welding of metallic materials are qualified according to ISO 14732 [11] standard. Standard is applicable for fusion welding processes (including beam technologies) and pressure welding.

Welding Operator—person who controls or adjusts any welding parameter for mechanized or automatic welding.

Weld setter—person who sets up welding equipment for mechanized or automatic welding.

The qualification test for welding operators and weld setters usually follows a preliminary welding procedure specification (pWPS) or welding procedure specification (WPS) prepared in accordance with the relevant part of ISO 15609 series [22–27]. Welding operators or weld setters are qualified by one of the following methods:

- qualification based on a welding procedure test in accordance with the relevant part of ISO 15614-1 [20].
- qualification based on a pre-production welding test in accordance with ISO 15613 [28].
- qualification based on a test piece in accordance with the relevant.

Provided that the welding operator or weld setter works according to a qualified WPS, there are no limitations on the range of qualifications other than those listed below.

- For automatic welding:
- change of the welding process (except variants within welding process 13 as defined in ISO 4063 [19]).
- welding with or without arc sensor and/or joint sensor.
- change from single-run-per-side technique to multi-run-per-side technique (but not vice versa).
- change of type of welding unit (including change in the robot control system).
- change from welding with arc sensor and/or joint sensor to welding without arc sensor and/or joint sensor (but not vice versa).
- For mechanized welding:
- change of the welding process (except variants within welding process 13 as defined in ISO 4063 [19]).
- change from direct visual control to remote visual control and vice versa.
- deletion of automatic arc length control.
- deletion of automatic joint tracking.
- addition of welding positions other than those already qualified in accordance with ISO 9606-1 [12].
- change from single-run-per-side technique to multi-run-per-side technique (but not vice versa).
- deletion of backing.
- deletion of consumable inserts.

The welding operator or weld setter qualification begins from the date of welding of the test piece(s), provided that the required testing has been carried out and the test results obtained were acceptable.

The welding operator or weld setter demonstrates functional knowledge appropriate to the welding unit combined with knowledge of welding technology.

Functional knowledge appropriate to the welding unit includes:

- welding sequences/procedures in the relevant process.
- joint preparation and weld representation in the relevant process.
- weld imperfections in the relevant process.
- welding operator's or weld setter's qualification.
- process operation (knowledge of programming (if relevant), control system and the signals given by this system, moving system, auxiliary equipment, jigs and fixtures and set-up, parameters and adjustments within the given procedures, safety regulations).

Knowledge of welding technology includes:

- requirements.
- welding equipment (types, essential components, identification, safety regulations, etc.).
- parent metals (identification, methods and control of pre-heating, control of interpass temperature).
- consumables (identification, storage, handling and conditioning of consumables, selection of correct size, cleanliness of wire electrodes and flux-cored electrodes, control of wire spooling, control and monitoring of gas flow rates and quality, principles of welding without consumables).
- safety and accident prevention (electrical risk, mechanical risk, risk of welding fumes and gases, noise risk, etc.).
- visual testing of welds.

1.1.5.4 Maintenance

General

Maintenance includes:

- routine maintenance,
- scheduled preventive maintenance,
- major overhauls.

The maintenance manager provides planning, implementation, improvement, and analysis of maintenance. Maintenance is carried out by the Maintenance manager service with the involvement (if necessary) of external providers.

Maintenance is carried out in accordance with the annual schedules:

- Annual routine maintenance schedule and preventive maintenance schedule—created for the QMS process.
- Annual overhauls schedule—created for the organization.

Annual maintenance schedules are created based on the equipment's technical requirements and the repair cycle data.

Maintenance is carried out according to the maintenance instructions. Maintenance Instructions include:

- equipment specifications,
- device description,
- equipment supervision and operation duties requirements,
- equipment supervision and operation rules during the shift,
- equipment inspection and repair procedure,
- safety.

When maintenance is performed, it is recorded in the equipment Maintenance Journal.

Equipment Factors that have a Major Effect on Welded Fabrication Quality

Welding technological process can be seen as the result of "energy source-heating source-product" system operation, reflecting the typical sequence of energy conversion to change the aggregate state of the product and welding materials (electrode or casing wire, flux, etc.).

For example, in the case of arc welding, the electrical energy supplied from the industrial network is converted by the power supply into the energy type that best meets the conditions of arc combustion. Moving the arc with the required speed along the connection line, in turn, provides an optimum thermal effect on the product and welding material (casing or electrode wire, flux). This, in turn, provides the formation of the welded joint.

The parameters of the technological process are closely related and the change of one of them can affect other and, consequently, the quality of the welded joint. Each welding process can be characterized by some number of generalized coordinates (parameters of welding process), between which there are certain connections. For example, there is a connection between current and voltage on arc, which are determined by the properties of power supply. However, the frequency of transfer of electrode metal droplets depends both on the current and voltage on the arc, and on the diameter and chemical composition of the electrode wire and the shielding gas.

The whole variety of parameters of the welding process can be divided into three groups:

(1) **energy**, characterizing the contribution of energy to the formation of welded joints.
(2) **kinematic** and **geometric**, characterizing the spatial position and displacement of the heating source relative to the product.
(3) **metallurgical**, including welding materials, characterizing the conditions of formation of the welded joint.

In production, the technological process of welding undergoes **disturbance**—uncontrolled influences which disturb normal course of process and lead to deviation of welded joint quality indicators from normative values. Disturbances can affect all parameters of the welding process without exception, but they can be classified more conveniently not by the application site, but by analogy with the parameters of the welding process.

The sources of energy and kinematic disturbances are the industrial network and own welding equipment: power sources, control equipment, electrode feed drive, displacement, etc.

The sources of metallurgical disturbances are contamination of materials, poor protection, uncontrolled changes in welding parameters.

In terms of organization of maintenance, it is appropriate to periodically control the main factors of equipment, which have a significant impact on welding fabrication quality assurance.

For all methods and all types of welding equipment, it is important to ensure the stability of the electrical resistance of the welding circuit. Accordingly, maintenance should include periodic tightening of all threaded or bayonet joints in the electrical circuit. For high-voltage welding equipment (inverter power supplies, oscillators, cathode-ray welding equipment), it is also necessary to periodically remove dust from conductive parts and appropriate insulators.

The wear of conductive elements can also cause the change of resistance of the welding circle. For example, conductive lugs for arc welding, change of electrodes geometry due to wear for contact spot welding.

The conformity of the manufacturer stated external volt-ampere characteristics of the welding power sources shall also be verified, with the simultaneous verification of current and voltage waveforms. A critical for the quality of a welded joint failure of the power semiconductor elements or the control system is often not possible to determine using analog or digital ammeters and voltmeters.

A separate factor in ensuring the stability of the electrical resistance of the welding circuit is the effectiveness of the grounding of the equipment. Grounding is not only a matter of electrical safety of the equipment. Disturbance in the ground circuit leads to frequent unpredictable failures in the work of automated and, in particular, robotic equipment.

The energy parameters of the arc welding process with the fusion electrode are also often affected by the electrode wire feed rate. The failure of the feed rate is associated with the failure of the control system of the corresponding engines of the welding unit or the slippage of the wire in the feed rollers.

Disturbance of the heating source movement speed leads to a change in the heat input and, consequently, adversely affects the quality of the welded joint technical maintenance should pay attention to the serviceability of the control system of the corresponding engines and possible slippage of the elements of the mechanical part of the movement system.

Another factor that can lead to defects is a malfunction of the equipment that fastens and fixes the welded workpieces in the set position. Possible causes are malfunction of the control system, wear of mechanical elements, improper clamping force due to malfunction of the pneumatic or hydraulic system. The maintenance procedure must include verification of the relevant components of the assembly equipment.

The disturbances associated with welding materials include the correctness of the choice and purity of the base metal, welding wire, shielding gases and flux. Contamination of materials leads to the appearance of most types of defects in welded joints (Sect. 3.1).

1.1.5.5 Marketing

Marketing—is the first process of product lifecycle (Fig. 1.9).

Marketing is an essential part of welded fabrication quality planning:

- requirements for characteristics of potential products that can satisfy the largest number of customers are defined based on market, competitor activities and market trends analysis,
- design proposals are developed based on the requirements for characteristics of potential products,

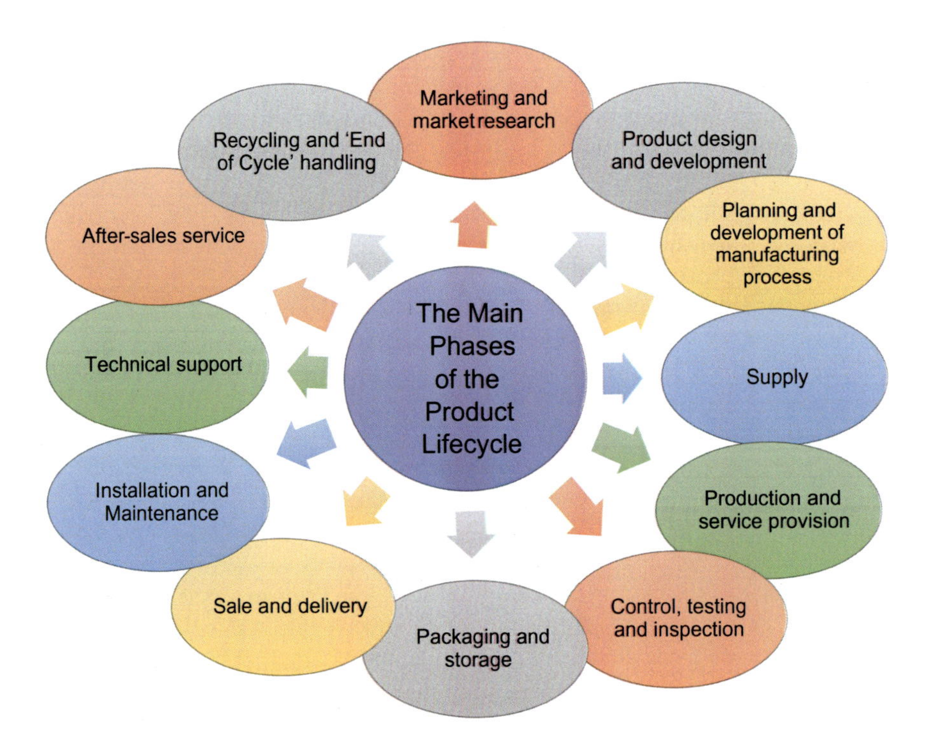

Fig. 1.9 Product lifecycle

- a degree of positive customer perception—the quality of potential products—is formed through advertising and promotion of products on the market,
- Through customer feedback, the real degree of product characteristics conformity to the customer requirements is evaluated—the quality of the products sold.

1.1.5.6 Contract Analysis

The main requirements of the ISO 9001 [2] standard for contract analysis are:

(1) Definition and recording customer requirements, as well as product use requirements and legislative and regulatory requirements in the contract.
(2) Analysis and ensuring the ability to meet all requirements prior to signing the contract.

The degree of conformity of product characteristics to the requirements (quality) can be high only if the requirements were defined.

'Contract Analysis' process is aimed at:

- determining requirements for products (services) when preparing the contract draft,
- analysis and ensuring meeting the requirements for products (services) before signing the contract,
- preservation of documented information about any new requirements analysis results,
- making amendments when the requirements for products (services) change.

1.1.5.7 Design and Development

Development and design of welding fabrication include:

- Knowledge and understanding of ISO 9001 [2] requirements for design and development (Sect. 1.5.7.1).
- Taking into consideration weldability—choice of welding materials, welding technologies, design of structure elements in a way which assures minimal influence of welding (Sect. 1.5.7.2).
- Assurance of testability of welded structure—design in respect to NDT (Sect. 1.5.7.3).
- Deciding on acceptable defects and imperfections and on desired quality level (Sect. 3.2).

ISO 9001 [2] Requirements for Design and Development

The earlier in the process the issues and errors are identified, the fewer resources are needed to detect and eliminate the causes.

Table 1.2 Typical product design and development stages

#	Stage	Stage output documented information
1	Specification development	'Specification' project
2	Specification verification (signing)	Approved 'Specification'
3	Draft project development	
4	Draft project verification (signing)	Draft project
5	Making a mockup	
6	Mockup testing	Measurement record
		Test report
7	Technical project development	
8	Technical project verification (signing)	Approved 'Technical project'
9	Manufacturing of a prototype (pilot batch)	
10	Validation (acceptance testing) of a prototype (pilot batch)	Prototype acceptance test report
11	Adjustment of design documentation based on the results of the prototype (pilot batch) acceptance testing	Design documentation (second edition)
12	First production batch production	
13	First production batch validation (testing)	First production batch test report
14	Adjustment of design documentation based on the results of the First production batch validation	Approved final edition of design documentation

To identify issues and errors in the early stages, the requirements of ISO 9001 [2] standard include the following:

(1) divide the design process into stages (ISO 9001 [2] does not regulate naming and content of stages),
(2) carry out verification and validation at appropriate stages,
(3) control and documentation changes made during, or after the design and development.

Typical product design and development stages are listed in Table 1.2.

Verify—confirm conformity to input requirements.

Validate—prove compliance with the terms of use (operation).

The concept of verification and validation in the 'Design and development' process is presented in Fig. 1.10.

Examples of verification and validation in the 'Design and Development process are presented in Fig. 1.11.

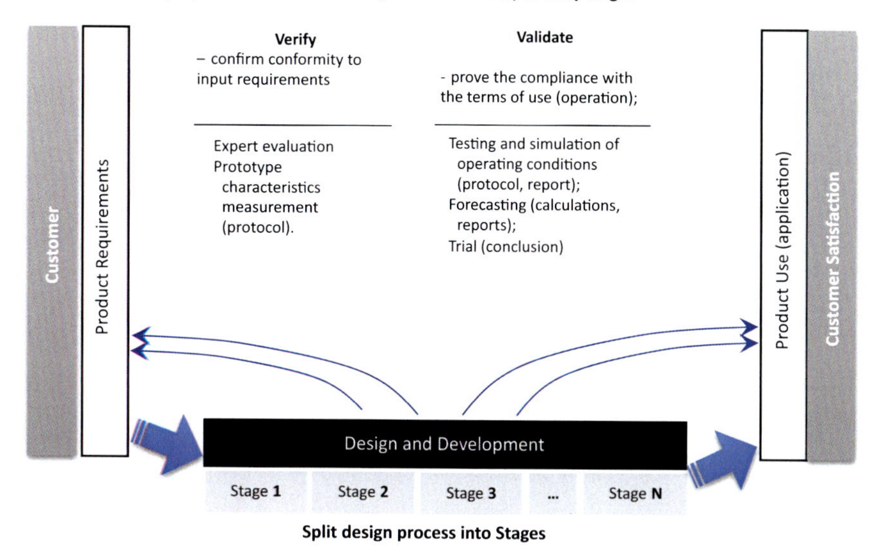

Fig. 1.10 The concepts of verification and validation in the 'Design and development' process

Weldability. ISO/TR 581 [29]

Weldability is a property of metal to form, with a pre-set welding technology applied, a non-detachable joint that satisfies requirements for metallurgical and mechanical characteristics of metal and assures suitability for the welded structure designation.

According to the Technical Report ISO/TR 581 [29] there are three factors influencing weldability (Fig. 1.12):

(1) material,
(2) technology of manufacturing,
(3) structure design.

(1) **Material** is related to **metallurgical weldability**—a property of metal (or metals) to form a monolith welded joint with acceptable structure. If the properties of welded joint are close to those of the base metal, metallurgical weldability is good.

Metallurgical weldability depends on chemical composition, metallurgical characteristics, and physical properties of the metal.

(a) Chemical composition is the main factor of metallurgical weldability.
(b) General rule for steels is as follows: increase of carbon and alloying elements content degrades metallurgical weldability.

Examples of verification and validation

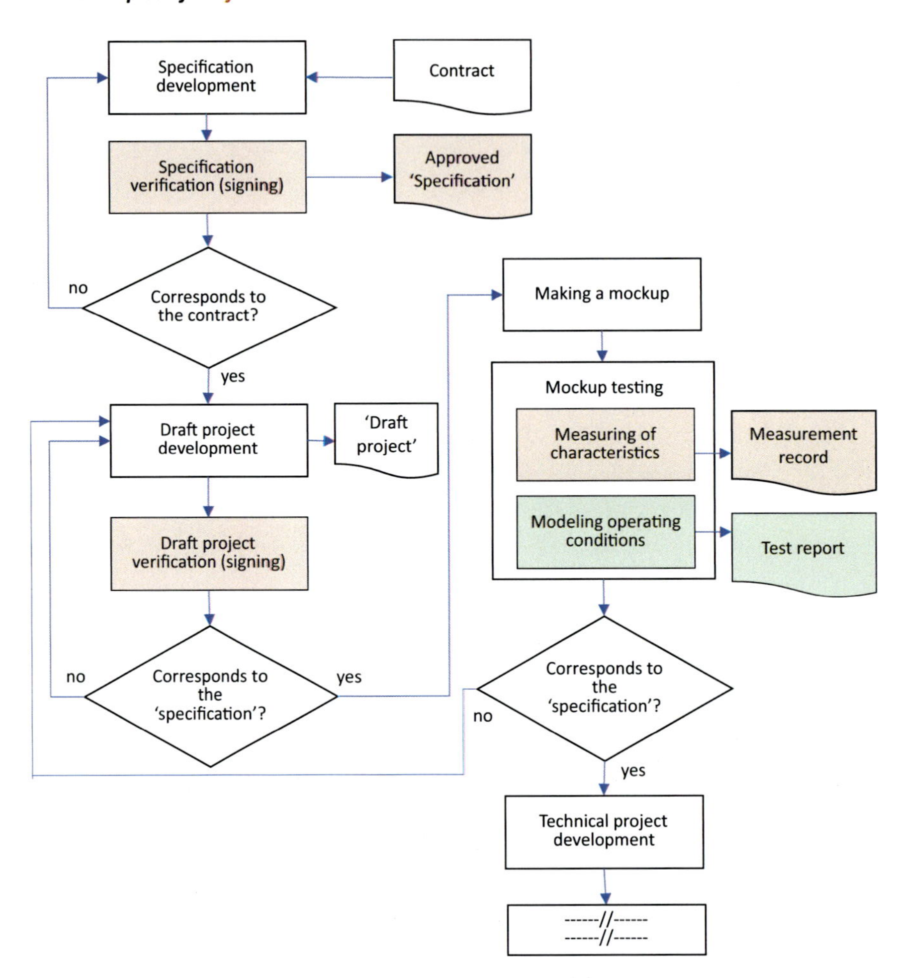

Fig. 1.11 Examples of verification and validation in the 'Design and development' process

(c) Metallurgical properties are defined by technology of foundry production, hot and cold treatment of parts including rolling and final heat-treatment.

Metallurgical characteristics influencing metallurgical weldability are:

- grain size and type of crystal structure—e.g., for steels the order of structures from best weldability to worst is ferrite > perlite > austenite > martensite,
- inclusions and segregations—impurities worsen weldability.
- Physical properties influencing metallurgical weldability are:
- tensile strength—weldability is worse for steels with high tensile strength,
- impact strength—weldability is better for steels and alloys with high impact strength,

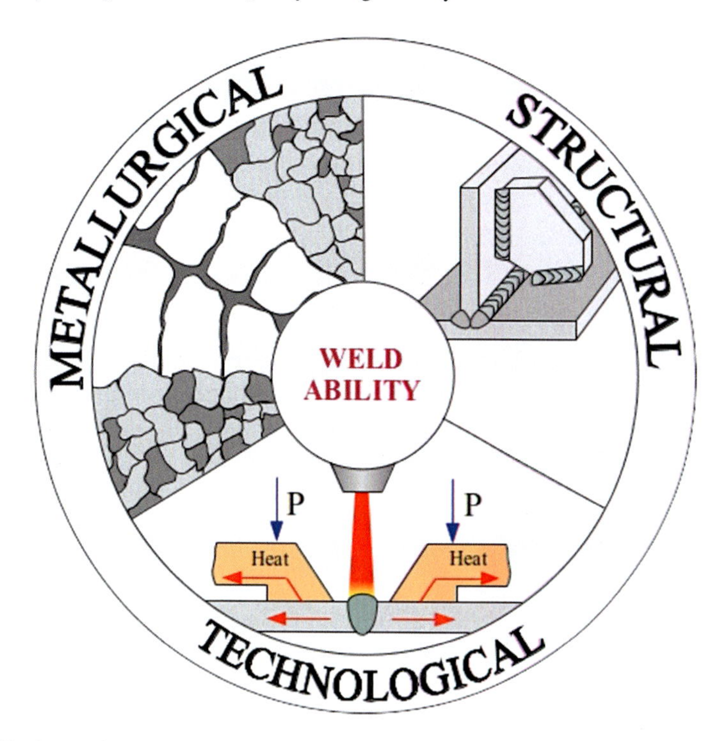

Fig. 1.12 Types of weldabilities of metals

- thermal conductivity—weldability is better for steels and alloys with high thermal conductivity,
- coefficient of temperature expansion—weldability is worse for steels and alloys with high coefficient of temperature expansion.

(2) **Technology of manufacturing** is related to **technological weldability**—property of the metal to react on a particular technology, mainly welding method, with formation of acceptable defects and imperfections (absence of cracks in the first place). Decrease of number of additional technological operations and equipment used (e.g., preheat, application of keyboard clamps for effective heat sink, etc.) improves technological weldability.

Technological weldability is ensured during the following stages:

(a) preparation for welding—by choosing joint type, edge preparation, fixtures, tack welds, etc.,
(b) welding itself—by choosing welding method, welding materials, welding parameters, position and sequence of welding, pre-heat, protection against environment-dependent factors, etc.,

(c) post-welding treatment—by choosing post-welding heat-treatment, mechanical treatment (e.g., grinding), chemical treatment (e.g., etching), etc.

(3) **Structure design** is related to **structural weldability**—rate of changes in metallurgical and technological weldabilities as a result of influence of structure elements on thermal cycle and formation of stressed state in different zones of welded joint. Decrease of number of structural features to be taken into consideration for the particular material and welding technology improves structural weldability of the structure.

Example—structural ribs:

- Increase stiffness of welded joint, which, in turn, decreases deformability of the weld and fusion zone during crystallization and cooling. This leads to an increase of residual stresses which contributes to formation of hot and cold cracks. On the other hand, the higher stiffness the structure possesses, the lower possibility of formation of defects with deviation of form and dimensions (Sect. 3.1, defects of group 5, subgroups 508 and 520).
- Increase heat sink near the rib, which, in turn, reduces the width of the heat-affected zone (HAZ). As a result, metal structure in the HAZ changes (usually improves) and shrinkage force reduces. As a result, the possibility of warping decreases (Sect. 3.1, defects of group 5, subgroup 520).

In the example given the structural weldability is defined by the combined effect of the factors listed.

Structural weldability depends on the following factors:

(a) Structure design—thickness of elements, mutual location and spatial position of welded joints, accessibility of welds, stiffeners,
(b) Operating conditions, working load at the first place—type of load (static, cyclic, impact), speed of load application, strength distribution, values of stresses, stress concentrators (notch effect), operating temperature, environment, and conditions of corrosion processes.

In practice materials in respect to weldability are usually divided into four groups. For steels, the criterion is carbon equivalent C_{eq} (Sect. 3.1.2):

(1) good weldability ($C_{eq} < 0.25$)—steel does not form cracks, requirements to welded joint are achieved without pre-heating and heating during welding as well as without pre- and post-welding heat-treatment,
(2) satisfactory weldability ($0.25 \leq C_{eq} < 0.45$)—steel does not form cracks, requirements to welded joint are achieved with pre-heating and post-welding heat-treatment,
(3) limited weldability ($0.45 \leq C_{eq} < 0.60$)—steel forms cracks, requirements to welded joint are achieved with pre-welding heat treatment, heating during welding and, as a rule, post-welding heat-treatment

(4) poor weldability ($C_{eq} \geq 0.60$)—steel forms cracks, requirements to welded joint are achieved with obligatory pre-welding heat treatment, heating during welding and post-welding heat-treatment.

Generally, to ensure structure design weldability, welding materials' choice and welding technology have to be developed so that the structure is able to operate in the defined conditions (load bearing capacity is at the required level) and to combine this with adequate security and minimum cost.

Design in Respect to NDT

An important task when designing the welded structure is to ensure its testability—fitness for being tested with a particular NDT method.

A good manufacturing practice is to set testability requirements as one of the inputs of design and development process, to include them into technical task for design and development or into manufacturer's standards.

Description of NDT system will be the corresponding output of design and development process. It should include three main sections:

(1) NDT methods and means (Chap. 4).
(2) Testing rules, including:
- details of work to be done during preparation of the structure for testing (e.g., removal of some parts to ensure accessibility),
- control points—zones of structure in which testing is to be performed,
- linking method characteristics—units where primary measuring or transitional devices are attached to the structure with details about linking methods, quality of surfaces to be linked and dimensions of linking zones,
- integrated measuring units—stationary systems with output of control signals to the external device,
- protection of control points, linking units and integrated measuring units from damage and pollution during structure operating,
- period and methods of testing,
- description of acceptable defects and imperfections,
- conditions and technology of removal of unacceptable defects,
- conditions of product disposal in case of unacceptable defects.

(3) Requirements to personnel (Sect. 1.5.3.4).

NDT system depends on structure quality level.

1.1.5.8 Control of Externally Provided Processes, Products, and Services

The main requirements of the ISO 9001 [2] standard for Control of externally provided processes, products and services are:

to determine and apply criteria for the evaluation, selection, monitoring of performance, and re-evaluation of external providers, based on their ability to provide processes or products and services in accordance with requirements.

'Control of externally provided processes, products and services' process (this process is often called 'Purchasing') ensures:

- formulating the supplier requirements,
- conducting a quantitative evaluation of suppliers at specified intervals,
- selecting suppliers by category, including the products procurement, outsourcing of processes, delivery on behalf of the organization,
- organization's QMS control over the externally provided processes,
- control of external providers as well as the resulting outputs,
- activities, necessary to ensure that the externally provided processes, products and services meet requirements.

Quality assurance in the welded fabrication in the "Control of externally provided processes, products and services" process:

- Ensures conformity of the characteristics of purchased metal, welding consumables, welding equipment with the design and development requirements.

1.1.5.9 Production and Service Provision

General

The main requirements of ISO 9001 [2] standard are:

(1) to implement production and service provision under controlled conditions,
(2) to provide traceability.

Quality assurance in welding manufacturing in the "Production and service provision" process is based on:

- controlled conditions of welding manufacturing, including validation of welding, heat treatment and coating,
- traceability,
- welding coordination.

Controlled Conditions

Controlled conditions include:

(a) The availability of documented information that defines:
  - the characteristics of the products to be produced, the services to be provided, or the activities to be performed (design and technological documentation),

- the results to be achieved (plans and programs of production and service provision).

(b) The availability and use of suitable monitoring and measuring resources. *Suitable* resources are defined as:
- resources exactly as they are determined in the technological documentation,
- resources should include validated or calibrated measuring equipment, approved measurement procedures, trained personnel conducting measurements.

(c) The implementation of monitoring and measuring activities at appropriate stages to verify that criteria for control of processes or outputs, and acceptance criteria for products and services, have been met. All product characteristics stated in technological documentation as well as all process parameters stated in technological documentation must be measured.

(d) The use of suitable infrastructure and environment for the operation of processes. *Suitable* means that infrastructure and environment should be exactly as defined in the documented information. At the same time, the equipment used must be serviced in a timely manner.

(e) The appointment of competent persons, including any required qualification. The competence of the personnel must meet the requirements of the relevant documented information.

(f) The validation, and periodic revalidation, of the ability to achieve planned results of the processes for production and service provision, where the resulting output cannot be verified by subsequent monitoring or measuring. Such processes are called special.

(g) The implementation of actions to prevent human error. An effective quality tool is poka-yoke.

(h) The implementation of release, delivery, and post-delivery activities.

Welding, Heat Treatment and Coating are Special Processes and Therefore Subject to Validation

Some examples of validation of welded structures:

- hydraulic tests of pipelines and pressure vessels under pressure exceeding operational values (usually by 1.25 or 1.5 times),
- tests of cranes and bridges under load exceeding operational conditions,
- crush tests,
- tests of models of welded structures in conditions simulating extreme situations (wind, earthquakes, waves, etc.),
- corrosion and strength tests (static, dynamic, fatigue) of samples of welded joints,
- mathematical simulation of influence of operating conditions on strength and resource of welded structures or their units.

Identification and Traceability

Identification—procedure of object recognition according to preset ID (marker).

Such markers may be bar-codes, labels with data, markings, etc.

Related to conditions of manufacturing, identification can be individual or sectional.

When individual identification is used, each part or unit should be marked directly. Individual identification is used in single and small-series production.

In case of sectional identification groups of materials, parts or units are marked. ID is usually located on the container (crate, bay wire, etc.). Sectional identification is used in mass production of simple structures.

Traceability—ability to follow the history of product creation (origin of materials and parts, technology and equipment used to manufacture each unit, personnel, shift, time and manufacturing conditions), location and operation of product on every part of its life cycle.

Traceability is vital for:

- analysis of causes of non-conformities and development of corrective actions.
- analysis of causes of positive effects and development of proposals for improvement.
- risk treatment.
- analysis of effectiveness of any actions taken (actions related to product manufacturing).

Traceability is assured by making notes on product in corresponding documents (chips, journals, route maps, corresponding tickets, etc.) with relation to product ID starting from materials up to packing and implementing. In this way unambiguous connection of two streams is ensured: the first being materials (raw materials, parts, units, etc.), the second being informational (electronic and hard copies of data about materials, parts, units, etc.)

Procedures for identification and traceability of products should be documented.

Aside from product life cycle traceability can be applied to calibrating of measuring devices. In this case traceability defines connection between measuring devices and National and/or international standards, standards of manufacturer, etc.

Brazing and Welding Procedure Specification (ISO 15607 [30] and ISO 15609 Series [22–27])

General rules of specification and qualification of welding procedures for metallic materials are given in the ISO 15607 [30].

Welding procedure—specified course of action to be followed in making a weld, including the welding process(es), reference to materials, welding consumables, preparation, preheating (if necessary), method and control of welding and post weld heat treatment (if relevant), and necessary equipment to be used.

Welding procedure specifications (WPS) provide a well-defined basis for planning of the welding operations and for quality control during welding.

Development and implementation of WPSs include the following phases.

1. **Development of the procedure**. At first WPS are developed as a preliminary document (pWPS) which should undergo qualification procedure prior to actual welding in production. pWPS contains the required variables of the welding procedure. It is prepared by the manufacturer using experience from previous productions and the general knowledge bank of welding technology. It should be absolutely applicable for the actual production.
2. **Qualification** is aimed at testing pWPS in production and to make changes (if necessary). This phase is performed by the manufacturer together with an examiner or examining body (if applicable). ISO 15607 [30] standard lists several methods for qualification. The manufacturer chooses one which, of course, should be the most suitable. Conditions and results of qualifications are documented as Welding procedure Qualification Record (WPQR). Each WPS should be qualified separately.

If the qualification involves welding of test pieces, those test pieces shall be welded in accordance with the pWPS.

The WPQR shall comprise all variables (essential and non-essential) as well as the specified ranges of qualifications.

3. **Finalization of procedure**. The result of this phase is a final version of WPS based on WPQR. WPS for production welding is developed under the responsibility of the manufacturer unless required otherwise.
4. **Release for production**—the manufacturer produces copies of WPS, or work instructions based on the final procedure to be used by the personnel.

ISO 15609 series [22–27] gives details of WPSs for different welding processes. It includes six parts:

- Part 1—arc welding [22].
- Part 2—gas welding [23].
- Part 3—electron-beam welding [24].
- Part 4—laser beam welding [25].
- Part 5—resistance welding [26].
- Part 6—laser-arc hybrid welding processes [27].

ISO 15609 series [22–27] defines procedures for metallic materials only.

WPS includes the following blocks of information:

(1) Related to manufacturer:

  - identifications of manufacturers.
  - identification of WPS.
  - reference to WPQR or other documents listed in ISO 15607 [30].

(2) Related to parent material:

- designation(s) with reference to the standard(s).
- group according to ISO/TR 15607 [30].
- dimensions (thickness, pipe diameter).
- dimensions (thickness, pipe diameter).

(3) Common for all welding processes:

- process(es) specified as listed in EN ISO 4063 [19].
- other information relevant to the group of processes (e.g., for arc welding one should specify type of weld, preparation of parts, cleaning prior to welding, welding position and technique, backing and method of its removal, filler materials and gas/flux, electric parameters of welding, traveling speed and wire feed speed for mechanized and automated processes, pre-heat and interpass temperatures, post-welding heat treatment details, heat input).

(4) Information specific to welding process (e.g., for TIG welding one should list tungsten electrode diameter and type, gas flow rate and nozzle diameter, additional filler materials).

Tolerance ranges shall be specified in the WPS for all variables.

Brazing and Welding Procedure Qualification (ISO 15610 [31], 156111 [32], 15612 [33], 15613 [34] and 15614 Series [20, 35–41])

Each pWPS shall be used as a basis for the establishment of Welding procedure Qualification Record (WPQR) qualified according to one of the following methods:

- Welding Procedure Test.
- tested welding consumables.
- previous welding experience.
- standard welding procedures.
- pre-production welding test.

A qualification is valid indefinitely for the ranges qualified unless specified otherwise.

Qualification based on welding procedure test specifies how a welding procedure can be qualified by the welding and testing of a standardized test piece.

A welding procedure test may be required whenever the properties of the material in the weld metal and the heat affected zone are critical for the application.

Requirements for welding procedure tests are given in ISO 15614 series:

Part 1—Arc and gas welding of steels and arc welding of nickel and nickel alloys [20].

Part 2—Arc welding of aluminum and aluminum alloys [35].

Part 3—Fusion welding of non-alloyed and low-alloyed cast irons [36].

Part 4—Finishing welding of aluminum castings [37].

Part 5—Arc welding of titanium, zirconium, and their alloys [38].
Part 6—Arc and gas welding of copper and its alloys [39].
Part 7—Overlay welding [40].
Part 8—Welding of tubes to tube-plate joints [41].
Welding procedure test includes:

- preparation of test pieces (material, thickness, and pipe diameter—according to pWPS, other dimensions—as listed in ISO 15614 series [20, 35–41]).
- Welding (parameters, welding materials, technique and sequence as listed in pWPS).
- testing (e.g., longitudinal weld tensile test, all weld metal bend test, corrosion test, chemical analysis, microscopic examination, delta ferrite examination, hardness test, cruciform test, impact test, non-destructive testing).

Qualification based on tested welding consumables specifies how a welding procedure can be qualified by using tested welding consumables. This method of qualification is described in the ISO 15610 [31] standard.

The standard applies to arc and gas welding, other fusion welding processes may be accepted if specified. ISO 15610 [31] is limited to application to parent metals that produce acceptable microstructures and properties in the heat affected zone and which do not deteriorate significantly in service.

However, this standard has limitations in application. It is *not* applicable where requirements for hardness or impact properties, preheating, controlled heat input, interpass temperature and post-weld heat-treatment are specified for the welded joint. The use of ISO 15610 [31] may also be restricted by an application standard or a specification.

Qualification based on previous welding experience specifies how a welding procedure can be qualified by demonstration of previous satisfactory welding ability.

A manufacturer may have a pWPS qualified by referring to previous experience in welding on a condition that they can prove that they have previously satisfactorily welded the type of joint and materials in question. The proof should be made by appropriate authentic documentation of independent nature. Only welding procedures known from experience to be reliable should be used in such cases. Qualification procedure is described in ISO 15611 [32] standard for the following processes: arc welding, gas welding, electron beam welding, laser beam welding, resistance welding.

Qualification by this method for other welding processes and for special applications may be also covered by specific standards (e.g., ISO 14555 [42] for stud welding or ISO 15620 [43] for friction welding). The use of ISO 15611 [32] may also be restricted by an application standard or a specification.

Qualification based on a standard welding procedure specifies how a welding procedure can be qualified by use of a standard welding procedure.

ISO 15612 [33] standard specifies how a user can follow a standard welding procedure specification (SWPS) based on welding procedure qualification tests performed by a different organization as well as requirements for qualification of welding procedures to be issued as SWPSs and the requirements for organizations adopting SWPSs.

This standard is applicable to welding of steels and aluminum and its alloys with arc welding, gas welding, electron beam welding, laser beam welding, resistance welding.

The use of ISO 15612 [33] can be restricted by an application standard or a specification.

Qualification based on a pre-production welding test specifies how a welding procedure can be qualified by pre-production welding tests.

This method is the only reliable method of qualification for some welding procedures. It is essential in cases when the resulting properties of the weld strongly depend on certain conditions. Such conditions may include component, special restraint conditions, heat sinks etc., which cannot be reproduced by standardized test pieces.

Qualification by a pre-production welding test may be used where the shape and dimensions of standardized pieces do not adequately represent the joint to be welded (e.g., attachment weld to thin pipe). In such cases, one or more special test pieces shall be made to simulate the production joint in all essential features. The test shall be carried out prior and under the conditions to be used in production.

Inspection and testing of the test piece shall be carried out in accordance with the appropriate standard for procedure testing, but this testing may need to be supplemented or replaced by special tests according to the nature of the joint in question.

ISO 15613 [34] defines the method of qualification by pre-production welding tests. It is applicable to arc welding, gas welding, beam welding, resistance welding, stud welding and friction welding of metallic materials.

The use of ISO 15613 [34] can be restricted by an application standard or specification.

1.1.5.10 Inspection

Control of Nonconforming Product

Inspection is a process of defining compliance of an object to the requirements.

Nonconformity—nonfulfillment of the requirement.

At the inspection process output products are divided into two groups—conforming and nonconforming.

Conforming products are delivered to the customer.

Nonconforming products must be controlled per ISO 9001 [2] requirements. Control of nonconforming products includes:

(1) Identification—labeling, for example with red paint, writing the word ‘defect’, corresponding bar code, etc.
(2) Segregation (containment). Defective product must be segregated in a way to exclude accidental use or nonconforming product delivery.

(3) Review of nonconformity and acceptance of one of three options:

- Correction of nonconformity (via removing the defect),
- Scrapping the product (if nonconformity cannot be removed),
- informing the customer and obtaining authorization for acceptance under concession (for example, discount pricing).

(4) Documenting the process describing:

- Nature of nonconformity (for example, imperfection type—Sect. 3.1, coordinates and size of the imperfection—Chap. 4, reference to the non-destructive testing protocol—Sect. 4.11, acceptance criteria—Sect. 3.2),
- Actions taken with respect to the nonconformity,
- Concessions obtained (if a customer was informed and authorization was obtained),
- Authorized personnel (or organization) who made decisions regarding the nonconformity.

Testing and Inspection Plan

Inspection and Test Plan (ITP) is a document detailing a systematic approach to testing a system or a product. It may include a variety of processes, e.g., visual inspection, dimension inspection, welding inspection, function test, factory acceptance test, etc. and participation of all parties.

ITPs are needed to guarantee the product quality and ensure the product function achieved the design requirement.

There are three major elements that should be described in the test plan:

- Test Coverage
- Test Methods
- Test Responsibilities.

Test coverage in the ITP states what requirements will be verified during which stages of the product life. Test Coverage is derived from design specifications and other requirements (safety standards, regulatory codes). Each requirement or specification of the design ideally should have one or more corresponding means of verification.

Test methods in the ITP state how test coverage will be implemented. Test methods may be determined by standards, regulatory agencies, contracts, etc. Test methods also specify test equipment to be used in test execution. Pass/fail criteria should be established as well.

Test responsibilities include which organizations will perform the test methods at each stage of the product lifecycle. This allows test organizations to plan, acquire or develop test equipment and other resources necessary to implement the test methods for which they are responsible. Test responsibilities also include information about data to be collected, and how that data will be stored and reported. The outcome of

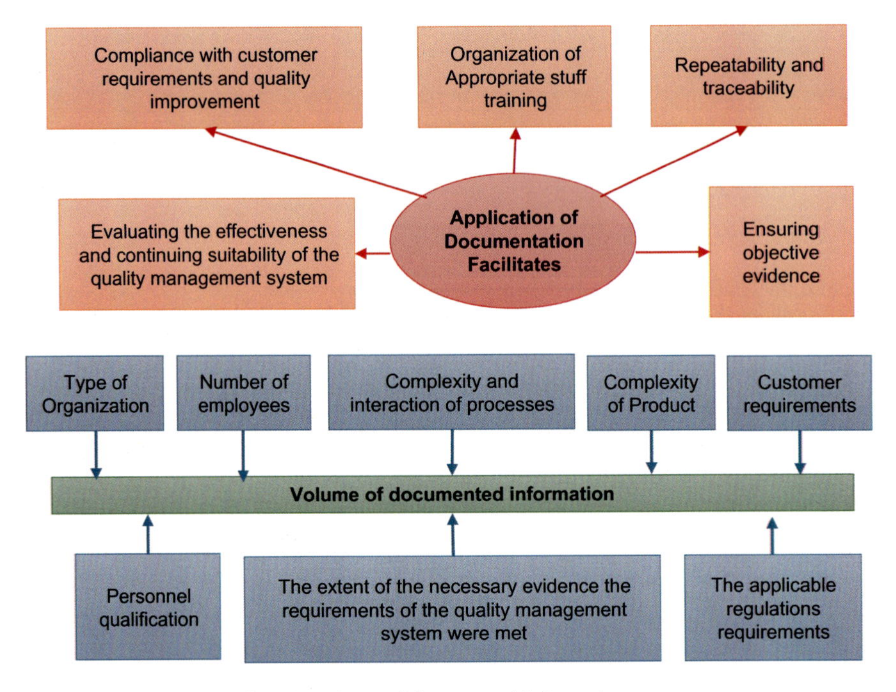

Fig. 1.13 Factors that define the volume of documented information

a successful test plan should be a record or report of the verification of all design specifications and requirements as agreed upon by all parties.

1.1.5.11 Control of Documented Information

General

Documented information—significant data and its medium (electronic or paper) that must be controlled.

The accepted practice is to divide documented information into two categories:

(1) **Specification**—documented information that establishes requirements (for example, standards, Quality Manual, Quality Plan, documented procedure, etc.).
(2) **Record**—documented information containing the results achieved or evidence of the activity performed (for example, reports, magazines, protocols, photographs, signatures, etc.).

Documented procedure—a description of how the activity or process is carried out—is an important kind of specification.

Volume of documented information is defined by (Fig. 1.13):

- **Type of organization**—the largest number of documents are owned by nuclear power and chemical industry enterprises,
- **Number of employees**—the larger the number of employees, the greater the amount of documented information,
- **Complexity of products and processes—The more complex the processes and products, the more** detailed and extensive the documented information that regulates them,
- **Customer and regulatory requirements**—can be associated with obligations for the availability of additional documented information,
- **Personnel competence**—less volume and detail of documented procedures is required for more competent staff,
- **Personnel training**—can replace the availability or the depth of documented procedures details,
- The extent of the necessary evidence the requirements of the Quality management system were met.

As per the recommendations of ISO 10013 [44], the organization chooses:

- Documenting method
- Terminology and type of documents, wherein it is important to ensure the unity of terminology in the QMS.
- Whether to make Quality Policy and Quality Objectives separate documents or include them in Quality Manual.
- The number of documented procedures—it is allowed to combine several procedures into one (for example, all documented procedures may be described in the Quality Manual).
- Structure (content) and Quality Manuals number
- Structure and form of the documented procedures

It is good practice when:

- Process owners and team members are the developers of documented information,
- Analogies, including organizational knowledge, are used in the development,
- Questionnaires and (or) user surveys are created to initiate the development of documented information,
- Personnel is trained in the control of documented information and new documentation,
- Validation is done—confirmation of documented information correctness by the practical implementation of edition 0,
- The process of documenting changes is simplified.

5W Method is used for documented information development. Documented information is considered self-sufficient if it answers the following 5 questions:

- ‘**W**’HAT to do (**W**HAT is done—for records)?
- ‘**W**’HO is doing it?—process owner, including responsibility and authority.
- HO‘**W**’ to do it?—sequence or the details of the activity.

- '**W**'HERE?—connection to Processes and (or) subdivisions.
- '**W**'HEN?—deadlines.

In accordance with ISO 9001 [2] in the QMS, mandatory documented procedures should be developed for the following processes:

- Control of documented information.
- Control of nonconforming outputs.
- Internal audit.
- Corrective action.

A documented procedure for managing the documented information should include creating and control.

Documented information creating includes:

- Identification document name, number, date, author
- Determination of format and document media
- Conducting verification analysis before the document is released.

Control of documented information includes:

- Ensuring availability of documented information at the workplace
- Protection of information, including confidentiality
- Distribution (retrieval)
- Ensuring storage and preservation
- Control of Changes
- Setting terms of disposition

Quality Manual

Quality Manual—the main specification that describes the functioning of the organization's quality management system.

Purpose of Quality Manual:

- Understanding the functioning of the organization's QMS as an object of control—for senior leadership and mid-level managers.
- Evaluation of the QMS effectiveness—for external and internal auditors, supervisors, senior leadership.
- Proof of the QMS effectiveness and the stability of the organization—for customers and partners.
- Quality management training—for internal auditors, senior leadership, mid-level managers.
- Organization's advertising—used in some cases by the sales department for potential customers.

As a rule, the contents of Quality Manual completely coincide with the contents of ISO 9001 [2]. The sections of Quality Manual describe how the organization fulfills the requirements of the corresponding section of ISO 9001 [2] and / or contains references to lower-level specifications, in which the requirements are described in more detail.

Quality plans. ISO 10005 [45]

Customers are often more interested in **how** the **quality** of the specific products they obtain is **ensured**, and **not** how the quality management system of the organization as a whole works. For these purposes Quality plan is used, as Quality Manual contains a lot of unnecessary information.

Quality Plan—specification that defines which procedures and related resources, when and by whom should be applied to a particular object.

Purpose of Quality Plan:

- To prove the effectiveness of ensuring a particular product quality—meeting customers, other interested parties, or the organization's own requirements.
- To evaluate and minimize the risk of not meeting requirements—for customers and mid-level managers.
- To develop quality assurance and quality control procedures for new product development—for the Design and Development department.
- In absence of the established QMS, Quality Plan is used for the Quality Manual tasks (see Sect. 1.5.11.2).

In terms of format, Quality Plan is a Mini Quality Manual, which describes the implementation of the processes of providing resources and the life cycle for specific products, as well as communication.

Guidelines (but not the mandatory requirements) for establishing, reviewing, accepting, applying, revising, and improving of Quality Plan are given in ISO 10005 [45].

ISO 10005 [45] standard contains guidelines for the following processes:

(1) Control of documented information
(2) Control of Resources:

  - Materials, products, and services,
  - People,
  - Infrastructure and environment for the operation of processes,
  - Monitoring and measuring resources.

(3) Interested party communication.
(4) Design and development
(5) Externally provided processes, products, and services
(6) Production and service provision
(7) Identification and traceability

(8) Property belonging to customers or external providers.
(9) Preservation of outputs
(10) Control of nonconforming outputs
(11) Monitoring and measurement
(12) Audits.

The requirements of ISO 10005 [45] for the listed processes comply with the requirements of ISO 9001 [2], addressed to specific products.

1.1.5.12 Internal Audit

Audit—systematic, independent, and documented process for obtaining audit evidence and evaluating it objectively to determine the extent to which the audit criteria are fulfilled.

In other words, an audit is a verification of compliance of what is actually being implemented with the established requirements (international standards, including ISO 9001 [2], specifications and etc.).

Audit evidence—objective data, obtained as a result of the audit.

Audit criteria—set of policies, documented information or requirements used as a reference against which audit evidence is compared (what compliance is verified against).

Audit program—activities to plan, conduct, monitor and improve the audits.

The main requirements of ISO 9001 [2] are:

- to plan an audit program,
- to define the audit criteria and scope for each audit,
- to take corrective actions without undue delay,
- to ensure that the results of the audits are reported and documented.

'Internal audit' process is aimed to achieving the following objectives:

(a) Provide information on whether the QMS conforms to:
    - the organization's own requirements for its QMS,
    - the requirements of International Standard ISO 9001 [2].
(b) Identify undesired effects and ensure their prevention or reduction.
(c) Identify desirable effects and ensure their enhancement.
(d) Provide information on whether the QMS is effectively implemented and maintained.

The Procedure of 'Internal Audit' process [46] includes (Fig. 1.14):

- Audits planning,
- Audit preparation and conducting,
- Results presentation and Audit Report,
- 'Internal Audit' process analysis and improvement

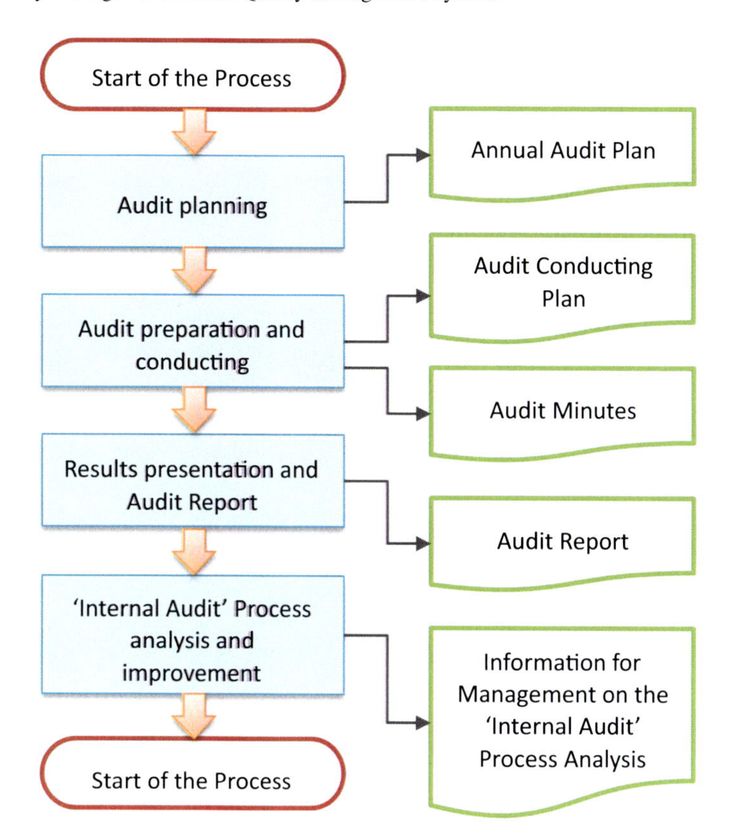

Fig. 1.14 'Internal audit' process diagram

Audit Planning

When planning audits Quality Manager:

- Draws the Annual Audit Plan,
- Forms the audit team and distributes the Lead Auditors responsibility and authority for Annual Audit Plan implementation,
- determines and provides the resources necessary to carry out the 'Internal Audit' Process, including the auditors training.

Auditor training must be verified by the internal auditor certificate or another similar training certificate.

Preparation and Conducting of the Audit

Lead Auditor, based on the Annual audit plan, creates an Audit conducting Plan for the specific audit. Audit conducting Plan points out which QMS process is verified, place of audit, audit checklist (ISO 9001 [2] clauses), Lead Auditor, audit team, date, and time. Lead Auditor distributes responsibility and authority among Audit team members.

One day before the audit start date, Lead Auditor requests the participating Process Owners to confirm the date and time of the audit (absence of obstacles).

Lead Auditor starts the audit with an opening meeting. The audit team and the inspected subdivisions' experts are presented at the meeting. At the opening meeting:

- The audit team is presented,
- Audit objectives are formulated,
- Attention is drawn to the selective nature of the audit,
- Information security and compliance with confidentiality requirements issues are specified,
- Communication during the audit is specified,
- Accompanying persons responsibility and authority are specified (if necessary),
- Audited subdivisions' experts' questions regarding conducting the audit are answered.
- When conducting the audit, the sample data is collected at the workplace via:
- Analysis of documented information,
- Interviewing,
- Monitoring the activities.

Verified sample data (the accuracy of which was confirmed) is the evidence of the audit. Auditor records evidence in the working audit journal and evaluates it to establish the extent of audit criteria fulfillment.

The effectiveness of implementation of corrective actions for nonconformities and recommendations for improvement identified during the previous audits are always checked.

The auditor reconciles the nonconformities and improvement recommendations identified during the audit with the Process Owner.

The audit ends with a closing meeting.

Before the closing meeting, the Lead Auditor gathers the audit team in order to:

- review the audit findings, and any other appropriate information collected during the audit, against the audit objectives,
- agree on the audit conclusions, taking into account the uncertainty inherent in the audit process,
- prepare recommendations.

The Lead Auditor holds the closing meeting in the presence of audit team and the experts of the audited units.

As appropriate, the following should be explained to the auditee in the closing meeting:

- advising that the audit evidence collected was based on a sample of the information available,
- the method of reporting,
- the process of handling of audit findings and possible consequences,
- presentation of the audit findings and conclusions in such a manner that they are understood and acknowledged by the auditee's,
- related post-audit activities (e.g., implementation of corrective actions).

Results Presentation and Audit Report

Based on the nonconformities and improvement recommendations reconciled with the Process Owner the Auditor draws Audit Minutes. Minutes are endorsed by the Process Owner (as a confirmation of the discrepancies wording agreement) and signed by the Lead Auditor.

Process Owner determines causes of nonconformities, develops corrective actions (or actions to implement improvement recommendations), sets the deadline for their implementation, and agrees with the Lead Auditor.

Corrective actions development period shall not exceed 3 business days (usually).

The original audit minutes are transferred to Lead Auditor to create the Audit Report, a copy of the Minutes is transferred to the Process Owner.

Lead Auditor or Auditor, who identified the nonconformity, checks the corrective actions fulfillment (or actions to implement improvement recommendations) within 7 business days (usually) since the implementation and adds a verification results records to the original and copy of the protocol.

Within 10 business days (usually), based on the audit, Lead Auditor creates Audit Report, which includes conclusions on the audit objectives and the original of Audit Minutes. Audit Report is approved by Quality Manager.

'Internal Audit' Process Analysis and Improvement

Quality Manager conducts 'Internal Audit' process analysis, which among all includes:

- Analysis and summary of the audit report,
- 'Internal Audit' Process criteria definition and analysis,
- Addressing risks and opportunities of 'Internal Audit' Process,
- Establishing trends for both undesired effects and desirable effects in the QMS processes.

Based on the analysis conducted, Quality Manager prepares information for the Management that includes:

(a) Information on whether the QMS conforms to:

- the organization's own requirements for its QMS,
- the requirements of International Standard ISO 9001 [2],

(b) Suggestions to prevent, or reduce, undesired effects,
(c) Suggestions to enhance desirable effects,
(d) Information on whether the QMS is effectively implemented and maintained,
(e) Suggestions on the Internal Audit program resources.

1.1.5.13 Corrective Action

Corrective action—an action taken to eliminate the causes of nonconformity and prevent its recurrence.

Correction—an action taken to eliminate the nonconformity (without analyzing its causes).

Corrective action.

ISO 9001 [2] requirements for actions in case of a nonconformity:

(1) React:

- correction,
- deal with the consequences.

(2) Evaluate the needed to eliminate the cause(s):

- reviewing and analyzing the nonconformity,
- determining the causes of the nonconformity,
- determining if similar nonconformities exist or could potentially occur.

(3) Implement the action needed.
(4) Review the effectiveness of any corrective action taken.
(5) Update risks and opportunities determined during planning (if necessary).
(6) Documented:

- the nature of the nonconformities and any actions taken,
- the results of any corrective action.

Corrective actions (CAR's) may be initiated when actual or potential nonconformities are discovered and apply to both products and processes. CAR's may be initiated:

- in the process 'Internal Audit',
- in any process of the QMS by any employee of the organization,
- for non-quality-related issues at the organization such as safety and cleanliness issues.

Product nonconformities relate to products sold by the organization, both produced in-house and purchased from vendors.

CARs are initiated using a CAR form. For 'Internal Audit' Process, a CAR Form is a part of the audit protocol.

CAR forms are given to the Process Owner who will review the request and determine if action is needed.

The Process Owner determines the root cause of the nonconformity, the corrective action needed to correct the problem, and the action needed to make sure it does not happen again. The Process Owner completes the appropriate sections of the CAR form.

Process Owner reviews the effectiveness of corrective action taken, conducts actualization of risks for the modified process and, if necessary, initiates QMS changes.

The results of any corrective action are recorded in the CAR forms. Completed CAR forms are kept on file by the Quality Manager.

Quality Manager prepares information on the results of analysis of nonconformities and corrective actions implementation in the QMS Processes for the Management Review.

1.1.6 Risk Management

1.1.6.1 General

Requirements for addressing risks and opportunities in decision-making in processes are included in all standards for management systems.

Risk is the impact of uncertainty on goal achievement.

Opportunity is the ability of an object to obtain an output that will correspond to the requirements for this output.

Opportunity is tied to the goal, and **risk**—to uncertainty when the goal is achieved.

To address risks in the QMS the following international standards can be applied:

- ISO GUIDE 73 [47]—Risk management. Vocabulary.
- ISO 31000 [48]—Risk management. Guidelines.
- IEC 31010 [49]—Risk management. Risk assessment techniques.

Make actions to address risks and opportunities a part of decision making on all levels of the organization and one of the improvement mechanisms of processes and quality management system.

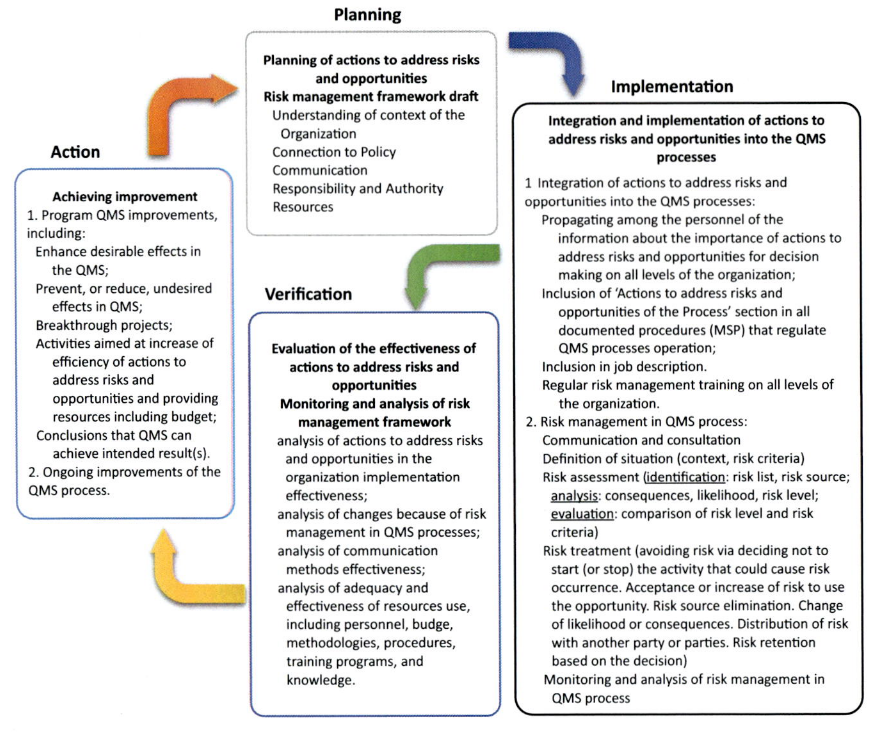

Fig. 1.15 Actions to address risks and opportunities diagram

Actions to address risks and opportunities (Fig. 1.15) consist of carrying out the 'PDCA' cycle and are implemented in four phases:

(1) Planning actions to address risks and opportunities. Project of risk management framework.
(2) Integration and implementation of actions to address risks and opportunities into QMS processes.
(3) Evaluation of the effectiveness of actions to address risks and opportunities. Monitoring and analysis of risk management framework.
(4) Achieving improvement

1.1.6.2 Planning Actions to Address Risks and Opportunities. Risk Management Framework Project

Understanding the Context of the organization is a prerequisite for the risk management framework project, defining the levels of risk and risk criteria, as well as risk treatment.

Information flows on internal and external context include:

- Information on the applicable laws ('Control of documented information' Process),
- Standards ('Control of documented information' Process),
- Information on markets and market trends ('Marketing Activity' Process),
- Information on competitors including technology ('Marketing Activity' Process),
- Information about the policy, objectives, strategies, promising business opportunities ('Management Review' Process),
- Information about the organizational structure and resources ('Management Review' and 'Personnel Management' Processes),
- Information about the technologies and future developments ('Design and Development' Process),
- Information about the personnel qualification ('Personnel Management' Process),
- Other information, specific to the case.

As means of communication in the 'Actions to address risks and opportunities' Process (including the flow of information about the internal and external context), training, meetings, internal information network (Intranet) are used.

Risk Management Principles are part of the Quality policy: Risk management is a part of the decision making. Effectiveness of risk management is based on the commitment of the leadership at all levels /of the organization/. Risk management is systematic, structured, coordinated in time, based on the best available information, and corresponds to the level of corporate culture.

Resources of 'Actions to address risks and opportunities' Process include:

- People (Leadership, Quality Manager, Risk Manager, Process Owners, Risk Owners, experts),
- 'Actions to address risks and opportunities' Process budget,
- Methods and procedures,
- Training programs,
- Organizational knowledge on risk management topics.

When calculating the level of risk, the simple formulas from Table 1.3 can be used.

Statistics of monitoring and measurement, and (or) expert evaluation (if data are insufficient or if they are absent) are used in calculations in the methods of risk management.

Expert evaluation is conducted by a team of experts consisting of 5 to 7 people. To facilitate the work of the experts, the assessment is carried out in qualitative categories, for example, in the case of determining the probability: 'very high'—'high'—'medium'—'low'—'very low'. It is possible to give recommendations what category is assigned under which conditions. Next the results are digitized. An example of the digitizing of the event probability expert evaluation results is given below (Table 1.4).

The maximum and minimum values are discarded, and the average value is determined from the remaining values—this is the result of the evaluation.

Table 1.3 Simple formulas for level of risk calculations

Risk calculating formula	Scope
$R = S \bullet O$	In quality management systems for the management processes
$R = S \bullet O \bullet D$	In technological processes to assess defects
$R = S \bullet O \bullet f$	In OSH management systems (ISO 45001:2017) to quantify the hazard

Definition	Value	Range	Method of determination
S	Consequences of the event	$1 \leq S \leq 10$	Expert evaluation
O	Probability of the event (in case of statistical evaluation, this is the probability p)	$1 \leq O \leq 10$ $0 \leq p \leq 1$	Expert evaluation,
D	Probability of detection (of the defect or its causes)	$1 \leq D \leq 10$	Expert evaluation
f	Exposure (the frequency with which the hazard is faced)	$1 \leq f \leq 10$	Expert evaluation

Table 1.4 Digitizing of the event probability expert evaluation results (example)

Expert evaluation	Evaluation recommendation (if appropriate)	Scale from 0 to 1	Scale from 1 to 10
Very high probability		1	10
High probability		0.75	7.5
Medium probability		0.5	5
Low probability		0.25	3
Very low probability		0	1

The results of actions to address risks and opportunities planning is the Risk Management Plan. Risk Management Plan is developed by Risk Manager.

1.1.6.3 Integration and Implementation of Actions to Address Risks and Opportunities into the QMS Processes

Integration of actions to address risks and opportunities into the QMS processes includes:

- Raising the personnel awareness of the actions to address risks and opportunities importance for the decision-making at all levels of the organization, as well as through the inclusion of this provision into the Quality Policy.

- The inclusion into the Quality Manager, Risk Manager, Process Owner job description of their responsibility and authority for the actions to address risks and opportunities.
- Conducting regular training on risk management at all levels of the organization.

Implementation of actions to address risks and opportunities (Risk management in QMS process) includes:

- Communication and consultation
- Situation determination
- Risk assessment (identification—analysis—evaluation)
- Risk treatment
- Monitoring and review of risk management process
- *Communication and consultation* between Risk Manager, team of experts, external consultants, representatives of the consumers and other interested parties are carried out for solving the following tasks:
- assistance in identifying and understanding the context for each specific assessment,
- ensuring that interested parties' interests are understood and considered,
- ensuring that risks are identified,
- bringing together the different areas of expertise for analyzing risks,
- determining risk criteria, assess risks and ensuring that different points of view are taken into account,
- ensuring confirmation and maintenance of the Risk Treatment Plan,
- ensuring change management in the 'Risk management process'.
- *Determination of the situation* is to analyze the context and defining the risk criteria.

In addition to the analysis of information flow on the external and internal context of the organization (described above) some internal and external conditions, in which the process is carried out, are determined:

- definition of goals and objectives for risk management activities in the given process,
- distribution of responsibility and authority in the 'Risk management process',
- definition of the scope, including the depth and breadth of activities of risk management,
- definition of activities, processes, functions, projects, products, services, or assets in respect of time and location,
- definition of the relationship between a specific project, process, or activity and other projects, processes, or activities in the organization,
- definition of risk assessment methodologies for the given process,
- definition of the method, by which the effectiveness of risk management will be assessed,
- identification and establishing the decisions to be made,

- identification, definition of the scope or the preparation of necessary studies and resources required for such studies.

While determining risk criteria consider the following factors:

- context, including legislation,
- the nature and type of the causes and consequences that may arise and how they will be measured,
- how the probability will be determined,
- timeframes of probability and /or consequences,
- how the level of risk will be determined,
- interests of interested parties,
- the level at which the risk becomes allowable or acceptable,
- whether a combination of multiple risks should be considered, and if so, how, and which combination.

Risk criteria determination is done via Expert Evaluation.

(c) *Risk assessment* consistent implementation of:
- *risk identification,*
- *risk analysis,*
- *risk evaluation.*

The goal of *risk identification*—is to compile a complete risk list, based on the risk cases, which may pave the way to increase the ability to prevent, degrade, and reduce the achievement of the objectives.

The source of risk is identified, as well as the area of its influence, risk events (including changes in circumstances), their causes, and their potential consequences Risks associated with lost opportunities are also identified.

Identification covers all risks (whether their source is under control of the organization or not), even if the source of the risk or its cause are not obvious.

Risk identification includes examining chain reaction of some specific effects, including a cascading effect and cumulative effect. It also examines a wide range of consequences, even if the source or cause of the risk is unclear. Along with the identification of the possible consequences, probable causes and scenarios that may indicate the alleged effects are considered.

When identifying risks, Cause and Effect diagrams and Brainstorming techniques are used.

Risk analysis consists of determining the level of risk for each risk listed in the risk list. Risk analysis is carried out by determining the consequences and their likelihood, as well as other characteristics relevant to the risk.

Consequences are determined via Expert Evaluation.

Likelihood is typically determined via Expert Evaluation, or if there is a database, using statistical methods.

Risk evaluation is an essential part of decision-making, and consists of the following:

- rankings—the arrangement of risks in risk list in the descending order of level of risk,
- comparison of the level of risk and risk criteria,
- allocation of risks in relation to which it is necessary to carry out the risk treatment.
- The goal of *Risk treatment* is to prevent, or reduce, undesired effects and to enhance desirable effects at the process level.

One or a combination of several options to address risks is used in order to prevent, or reduce, undesired effects:

- avoiding risk,
- taking risks to pursue an opportunity,
- eliminating the risk source,
- changing the likelihood or consequences,
- sharing the risk or retaining risk by informed decision.
- To enhance desirable effects and expand opportunities the following actions are taken:
- adoption of new practices,
- launching new products,
- opening new markets,
- addressing new clients,
- building partnerships,
- using new technology,
- other desirable and viable possibilities to address the organizations or its customers' needs.
- *Monitoring and review of risk management process* includes:
- verification and evaluation of Risk Treatment Plan implementation effectiveness,
- management methods adequacy analysis,
- obtaining additional information in order to improve risk assessment,
- analysis and lessons learnt from the risky of cases (including accidents), changes, flow, success, and failures,
- detection of changes in the external and internal context, including changes in risk criteria and the risk itself, that may require verification of risk treatment and priorities,
- identification of new emerging risks.

1.1.6.4 Evaluation of Effectiveness of Actions to Address Risks and Opportunities

The evaluation of the effectiveness of actions to address risks and opportunities consists of:

- analysis of Risk Management Plan implementation effectiveness,
- analysis of changes in the QMS processes because of actions to address risks and opportunities,
- determination of the values and analysis of the criteria changes in the 'Actions to address risks and opportunities' Process,
- analysis of effectiveness of communication methods and reporting mechanisms,
- analysis of the adequacy and effectiveness of resource use, including people, budget, methods and procedures, training programs, and knowledge.

Risk Manager prepares:

(a) current reports to Quality Manager on the results of addressing risks and opportunities in the QMS Processes, including:
   - suggestions for the QMS processes improvement,
   - information on the effectiveness of the 'Actions to address risks and opportunities' Process for Management review,
   - suggestions for improvement and actions for risk treatment, which require additional resources and (or) associated with multiple processes and require QMS changes,

(b) Annual report to the Company's management on the results of the Risk Management Plan and proposals to improve the efficiency of the 'Actions to address risks and opportunities' Process.

1.1.6.5 Achieving Improvement.

Improvement in 'Actions to address risks and opportunities' Process is achieved by two vehicles.

(1) Program QMS improvement that, among others, include:
   - enhancing desirable effects in the QMS,
   - preventing, or reducing, undesired effects in the QMS,
   - breakthrough projects,
   - measures aimed at increasing efficiency of the 'Actions to address risks and opportunities' process and resource provision, including budget,
   - conclusions that QMS can achieve its intended result(s).

(2) Current improvements to the given QMS process are carried out by the Process Owner:
   - as a part of the Risk Treatment Plan implementation,
   - based on improvement proposals analysis that result from the Risk Owner report on the results of risk management process monitoring and review.

References

1. ISO 9000: 2015 Quality management systems—Fundamentals and vocabulary
2. ISO 9001: 2015 Quality management systems—Requirements
3. ISO 3834:2005 Quality requirements for fusion welding of metallic materials
4. ISO 3834-1:2005 Quality requirements for fusion welding of metallic materials—Part 1: Criteria for the selection of the appropriate level of quality requirements
5. ISO 3834-2:2005 Quality requirements for fusion welding of metallic materials—Part 2: Comprehensive quality requirements
6. ISO 3834–3:1994 Quality requirements for fusion welding of metallic materials—Part 3: Standard quality requirements
7. ISO 3834-4:2005 Quality requirements for fusion welding of metallic materials—Part 4: Elementary quality requirements
8. ISO/TC 176/SC 2/N 544R Secretariat of ISO/TC 176/SC 2 ISO 9000 Introduction and Support Package: Guidance on the Process Approach to quality management systems
9. BS 6143-1—1992 Guide to the Economics of Quality—Part 1 Process Cost Model PDF
10. ISO 14731:2006 Welding coordination. Tasks and responsibilities
11. ISO 14732:2013 Welding personnel—Qualification testing of welding operators and weld setters for mechanized and automatic welding of metallic materials
12. ISO 9606-1:2012 Qualification testing of welders—Fusion welding—Part 1: Steels
13. ISO 9606-2:2004 Qualification test of welders—Fusion welding—Part 2: Aluminum and aluminum alloys
14. ISO 9606-3:1999 Approval testing of welders—Fusion welding—Part 3: Copper and copper alloys
15. ISO 9606-4:1999 Approval testing of welders—Fusion welding—Part 4: Nickel and nickel alloys
16. ISO 9606-5:2000 Approval testing of welders—Fusion welding—Part 5: Titanium and titanium alloys, zirconium, and zirconium alloys
17. ISO 9712:2021 Non-destructive testing—Qualification and certification of NDT personnel
18. IIW Guideline for International Welding Engineers, Technologists, Specialists and Practitioners. Personnel with qualification for welding coordination. Minimum requirements for education, training, examination, and qualification
19. ISO 4063:2009 Welding and allied processes—Nomenclature of processes and reference numbers
20. ISO 15614-1 (Part 1):2017 Specification and qualification of welding procedures for metallic materials—Welding procedure test—Part 1: Arc and gas welding of steels and arc welding of nickel and nickel alloys
21. ISO/TR 15608:2017 Welding—Guidelines for a metallic material grouping system
22. ISO 15609-1:2019 Specification and qualification of welding procedures for metallic materials—Welding procedure specification—Part 1: Arc welding
23. ISO 15609-2:2019 Specification and qualification of welding procedures for metallic materials—Welding procedure specification—Part 2: Gas welding
24. ISO 15609-3:2004 Specification and qualification of welding procedures for metallic materials—Welding procedure specification—Part 3: Electron beam welding
25. ISO 15609-4:2009 Specification and qualification of welding procedures for metallic materials—Welding procedure specification—Part 4: Laser beam welding
26. ISO 15609-5:2011 Specification and qualification of welding procedures for metallic materials—Welding procedure specification—Part 5: Resistance welding
27. ISO 15609-6:2013 Specification and qualification of welding procedures for metallic materials—Welding procedure specification—Part 6: Laser-arc hybrid welding
28. ISO 15613:2004 Specification and qualification of welding procedures for metallic materials—Qualification based on pre-production welding test
29. ISO/TR 581:2005 Weldability—Metallic materials—General principles

30. ISO 15607:2003 Specification and qualification of welding procedures for metallic materials—General rules
31. ISO 15610:2023 Specification and qualification of welding procedures for metallic materials—Qualification based on tested welding consumables
32. ISO/DIS 15611 Specification and qualification of welding procedures for metallic materials—Qualification based on previous welding experience
33. ISO 15612:2004 Specification and qualification of welding procedures for metallic materials—Qualification by adoption of a standard welding procedure
34. ISO/DIS 15613 Specification and qualification of welding procedures for metallic materials—Qualification based on a pre-production welding test
35. ISO 15614-2 (Part 2):2005 Specification and qualification of welding procedures for metallic materials—Welding procedure test—Part 2: Arc welding of aluminum and its alloys
36. ISO 15614-3:2008 Specification and qualification of welding procedures for metallic materials—Welding procedure test—Part 3: Fusion welding of non-alloyed and low-alloyed cast irons
37. ISO 15614-4:2005 Specification and qualification of welding procedures for metallic materials—Welding procedure test—Part 4: Finishing welding of aluminum castings
38. ISO 15614-5:2004 Specification and qualification of welding procedures for metallic materials—Welding procedure test—Part 5: Arc welding of titanium, zirconium and their alloys
39. ISO 15614-6:2006 Specification and qualification of welding procedures for metallic materials—Welding procedure test—Part 6: Arc and gas welding of copper and its alloys
40. ISO 15614-7:2016 Specification and qualification of welding procedures for metallic materials—Welding procedure test—Part 7: Overlay welding
41. ISO 15614-8:2016 Specification and qualification of welding procedures for metallic materials—Welding procedure test—Part 8: Welding of tubes to tube-plate joints
42. ISO 14555:2017 Welding—Arc stud welding of metallic materials
43. ISO 15620:2019 Welding—Friction welding of metallic materials
44. ISO 10013:2021 Quality management systems—Guidance for documented information
45. ISO 10005:2018 Quality management—Guidelines for quality plans
46. ISO 19011:2018 Guidelines for auditing management systems
47. ISO Guide 73:2009 Risk management—Vocabulary
48. ISO 31000:2018 Risk management—Guidelines
49. IEC 31010:2019 Risk management—Risk assessment techniques

Chapter 2
Measurement, Control and Recording in Welding

2.1 Measurement

The parameters of the welded fabrication process are determined by the relevant technical welding procedure specifications (WPS). All parameters are quantitative and must be specified in accordance with current international and national standards in WPS in certain values. Metrologists define [1] the concept of quantity as a property of the phenomenon, body, or substance, when this property has a size that can be expressed as a number indicating a distinctive feature. Such properties are, for example, the dimensions of the product in meters (units of length, distance) or the weight of the structure in kilograms.

During the adjustment of welding equipment and the course of the technological process and quality control of the structure, certain values are subject to measurement. Measurement is the process of experimentally obtaining one or more values of a quantity that can reasonably be attributed to the quantity. This process combines a measuring device, an operator performing the process, and the measurement technique—a detailed description of the measurement procedure together with the necessary calculations. Measurement only makes sense if a certain community uses the same system of measured values and standardized measurement procedures. The system of quantities is a set of quantities together with a set of consistent equations that connect these quantities.

The obtained measurement results are most often presented in the metric international system of physical units (SI) which is a set of standards for their determination. Standards are not constant and change according to the current level of science and technology. From May 20, 2019, all SI units are determined based on physical phenomena. That is, today, the basic units of SI are not determined by certain properties of man-made objects.

The international system of units is based on seven basic values: length (meter, m), mass (kilogram, kg), time (second, s), electric current (ampere, A), thermodynamic temperature (kelvin, K), amount of substance (mole), light power (candela, cd).

S. Fomichov et al., *Quality Management in Welded Fabrication*,
https://doi.org/10.1007/978-3-031-34800-6_2

Along with the SI system in Great Britain, the United States and other countries, the English Imperial system of measures officially adopted by the British Empire in 1824 is still used. Today, the measures of the English Imperial system are gradually being replaced by the metric international system of physical units (SI).

Base quantity is the value of a subset conditionally selected for a given system of quantities so that none of the values of the subset can be expressed in other quantities.

Together with the basic values, derived quantities are used—values that are defined in the system of quantities through the basic values of this system (m/s—unit of speed, m^2 kg s^{-3}—unit of power, watts).

2.1.1 Measurement Accuracy

Measurement accuracy—the closeness between the measured value and the true value of the measured value. Measurement accuracy is not an assessment of the accuracy of measurements.

The true value of the measured physical quantity, as a rule, is not known to us and it is impossible to obtain it. Therefore, in metrology, the concept of reference value of a quantity is used for practical application instead of the true value. **Reference quantity value** is a value of a quantity used as a basis for comparison with a value of the same kind. The reference value can be defined in different ways (as a method, the result of comparison with the standard, etc.) but, in any case, has an official status.

The accuracy of measurements is determined by many factors—measurement conditions, equipment, personnel performing measurements, etc. Numerical estimation of accuracy is a **measurement error**—the difference between the measured value and the reference value.

Errors of measuring instruments or, in other words, ***instrumental errors***, arise due to the imperfection of the design of the measuring instrument, as well as due to the ultimate capabilities of the technology of its manufacture.

The difference between the measured value and its reference value can be determined in different ways.

Absolute measurement error is defined as the difference between the measurement result and the reference value of the measured value. The error in this case is expressed in absolute units of the measured value.

Often the value of the absolute error does not give a correct idea of the accuracy of the measurement. For example, is the error in measuring a distance of 1 mm large or small? It is impossible to give an unambiguous answer to this question without estimating the reference value of the measured value. The error of 1 mm when measuring a distance of 5 mm will be huge, but when measuring a distance of 10 km—negligible. It is for such cases that the concept of relative error is used.

Relative measurement error—the ratio of the absolute measurement error to the actual or reference value of the measured value:

$$\delta = \frac{x - x_{ref}}{x_{ref}} \quad (2.1)$$

There can be many reasons for the error. Errors are classified as:

- instrumental error—due to the means of measuring equipment.
- methodological errors—due to inconsistency of the accepted model of object measurement.
- subjective errors—due to the qualifications, physical and mental condition of the operator.

Errors are also manifested in different ways. On this basis, there are:

- random errors—due to the result of the simultaneous impact on the measurement process of many unpredictable independent factors, each of which individually has little effect on the measurement result.
- systematic errors—due to the characteristics of measuring devices, methods or conditions of measurement and change over time according to a certain law.
- drift errors—due to the characteristics of measuring devices, methods or conditions of measurement and change over time in an unpredictable way.
- gross errors—due to the result of the impact on the measurement process of one or more random factors that significantly affect the measurement result.

Depending on the conditions under which the measuring instrument is operated, a distinction is made between basic (for normal conditions) and additional (if one or more influential values are outside of normal conditions) errors.

The main error is the error of the measuring instrument under normal conditions of its use.

Additional error—an error of the measuring instrument which additionally arises during the use of the measuring instrument in the conditions of deviation of at least one of the influential values from the normal value or its going beyond the normal range of values.

To estimate in advance the error that will be made by a particular equipment in the final result, the normalized error values are used.

The normalized values are the errors that are the limit for the certain type of measuring instruments. The standards regulate the methods of rationing and forms of expression of permissible error limits.

A generalized characteristic of a measuring instrument is an accuracy class—a classification characteristic of measuring instruments or measuring systems that meet the established metrological requirements, which are designed to maintain measurement errors or instrumental uncertainties within specified limits under certain operating conditions.

The accuracy class characterizes the accuracy of the measuring instrument, but it is not a direct characteristic of the accuracy of the measurement performed with this measuring instrument. The assignment of the accuracy class is based on the main error of the measuring instrument and the method of its expression.

The total measurement error can be reduced by introducing corrections to compensate for systematic errors. However, even in this case, there is no reason to believe that the result obtained corresponds exactly to the value of the measured value because it is impossible to estimate all the random effects on the measurement process. That is, the measurement result is only an estimate of the value of the measured value and should be supplemented by an estimate of its uncertainty, a quantitative expression of the value of doubt in the measurement results.

Measurement uncertainty is a non-negative parameter that characterizes the scattering of values that are attributed to a measured value and are based on the applicable information. This parameter is the estimate of the standard deviation.

Measurement uncertainty of type A is estimated based on statistical analysis of a number of results of repeated measurements of the same value.

Measurement uncertainty of type B is estimated based on non-statistical information—the design of the measuring instrument, the conditions under which the measurement is performed and more. Uncertainty of measurement in type B is estimated, as a rule, at impossibility of carrying out of repeated measurements.

The classification of measurement uncertainty by types A and B is introduced only to indicate different ways of determining the probability distribution of measurement results as random variables. Estimation of measurement uncertainty in any case is carried out by variance and standard deviation.

2.1.2 Unity of Measurement

Unity of measurements—the state of measurements, in which their results are expressed in units of measurement defined by law, and the characteristics of errors or uncertainties of measurements are known with a certain probability and do not go beyond the established limits.

Unity of measurements is achieved based on unity:

- standards and measures.
- calibration of metrological characteristics of measuring instruments.
- methods of measuring processes.
- forms of presentation of measurement results.

It is possible to achieve unity of measurements only at the stated level by creation of the corresponding legislative base and the state metrological service. Many countries have adopted laws governing metrological activities. For example, in Ukraine there is a law "On metrology and metrological activities."

State tests, metrological certification, verification, and calibration of measuring instruments are types of metrological control and supervision over the development, production, and operation of measuring instruments. These services are provided by the bodies of state and other metrological services, and this is a kind of lever of influence on business entities that use measuring equipment in their activities. The current legislation on metrology and metrological activities applies to all business entities, regardless of ownership.

Verification of measuring instruments (verification)—providing objective evidence that the object fully meets the established requirements. The verification includes the verification and marking and/or issuance of documents on the verification of the measuring instrument, which establish and confirm that the specified instrument meets the established requirements. National legislation defines the organizations authorized to carry out the verification of measuring instruments that are subject to the legislation on metrology and metrological activities.

There are primary and periodic verifications:

- primary verification of measuring instruments—verification of measuring instruments that have not been verified before, for example, immediately after manufacture.
- periodic verification of measuring instruments—verification carried out during the period of operation of measuring instruments after a set period of time (inter-calibration interval).

If the measurements are subject to the legislation on metrology and metrological activities, the relevant measuring instruments may be used, released from production, repair and sale, or rented only if they comply with laws and other regulations containing requirements for such measuring instruments.

Conformity assessment of legally regulated measuring instruments is carried out in the case when it is provided by the relevant technical regulations. A certificate is issued based on the results of the conformity assessment.

Types of metrological supervision are:

- state market supervision over the compliance of legally regulated measuring instruments with the requirements of technical regulations.
- metrological supervision of legally regulated measuring instruments in operation.
- metrological supervision over the quantity of packaged goods in packages.

In the presence of relevant international agreements, the results of conformity assessment, verification and calibration of measuring instruments carried out in other countries are recognized.

The International Organization of Legal Metrology (OIML) is responsible for the harmonization of national regulatory acts and metrological control procedures. The activities of this organization are aimed at the legal provision of the world economy with an efficient metrological infrastructure.

As of April 2020, the OIML has 61 member countries and 62 observer countries.

2.1.3 Measurement Classification

The classification of measurements can be carried out according to various criteria and, like any classification, is, to some extent, conditional.

According to the characteristics of measurement, accuracy is divided into:

- equal accuracy—a set of measurements performed by measuring instruments of the same accuracy and under the same conditions.
- non-equal accuracy—a set of measurements performed by measuring instruments of different accuracy or in different conditions.

According to the number of measurements, measurements are divided into:

- single measurements—a measurement of a physical quantity that is made once.
- multiple measurements—measurement of a physical quantity done several times (not less than four) to reduce the error and determine the result as the arithmetic mean of the measurements.

By type of measurement, values are divided into:

- static measurements—measurement of a physical quantity constant in time.
- dynamic measurements—measurement of a physical quantity that changes over time.

According to the purpose of measurement, measurements are divided into:

- Metrological Measurements—measurements of physical quantity performed using standards.
- technical measurements—measurements of a physical quantity performed using technical means of measurement.

According to the methods of presenting the measurement results, measurements are divided into:

- absolute measurements—measurements that are performed by direct measurement of basic quantities and (or) the use of physical constants.
- relative measurements—measurements in which the ratio of the measured physical quantity to a homogeneous basic quantity (unit) is determined.

According to the methods of obtaining measurement results, measurements are divided into:

- direct measurements—measurements of a physical quantity made by comparison directly with its measure (for example, the length of a weld is compared with a measure—a ruler).
- non-direct measurements—compatible measurements—measurements in which the value of the desired measure is determined on the basis of knowledge linking the measured value with a set of some direct measurements of heterogeneous quantities (for example, measuring the density of material part by mass and geometric parameters).
- aggregate measurements—measurements of a physical quantity, in which the values of the measured quantities are determined by repeated measurements of one or more homogeneous quantities at different combinations of measures or

these quantities (used, for example, to calibrate a set of measures according to the standard).

- compatible measurements are measurements in which at least two inhomogeneous physical quantities are measured to determine the functional relationship between them (for example, the electrical resistance of a conductor to temperature).

Measurements based on the peculiarities of determining their errors are divided into:

- laboratory measurements—are performed by highly qualified specialists most often with the use of universal exemplary measuring instruments in scientific experiments, in metrological research of unit standards and in the development and certification of methods of technical measurements. These measurements include the analysis of errors of measurement results.
- Technical measurements are performed by specialists whose duties do not include the analysis of errors in measurement results. Such measurements are usually used in production.

2.1.4 Parameters of Measured Values

When it comes to a measurand that changes over time, it is necessary to consider both its physical nature and what characteristics of the quantity to be measured. In the practice of measurements, it is established to distinguish the following characteristics of measured values: instantaneous value, peak or amplitude value, mean value, root mean square (effective value), rectified mean value.

Instantaneous value—the value of the measured value at a certain point in time. The instantaneous value corresponds to the ordinate of any point on the curve shown in Fig. 2.1 Instant value is rigidly related to time. That is, the instantaneous value of X is x_1 at time t_1 and x_2 at time t_2.

Peak (amplitude) value—the largest instantaneous value of the measured value for the period. In Fig. 2.1 the peak value corresponds to the value of x_a. It should be noted that the amplitude value x_a of the value of X is the instantaneous value at time t_4.

The average value is a constant component of the measured value. For a continuous quantity (as in Fig. 2.1), the average value is calculated for a certain period t_1-t_0 as follows:

$$\delta = \frac{x - x_{ref}}{x_{ref}} \tag{2.2}$$

For a sinusoidal value, the average value for the period is zero.

The RMS value of a continuous measurand (otherwise effective—the term came from electrical engineering) is calculated for the period of change of T by the formula:

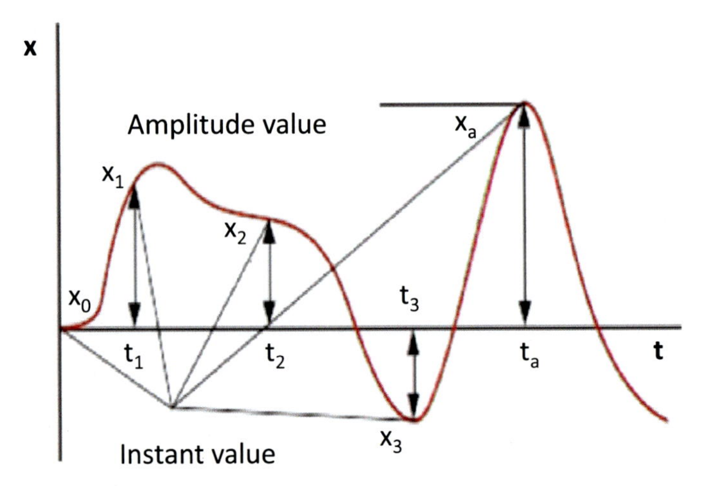

Fig. 2.1 Analog signal parameters

$$x' = \frac{1}{(t_1 - t_2)} \int_{t_1}^{t_2} x(t)dt \tag{2.3}$$

Average rectified value (another term in electrical engineering)—the average value of the modulus of the measured value

$$s = \frac{1}{(t_2 - t_1)} \cdot \int_{t_1}^{t_1} |x(t)|dt \tag{2.4}$$

The decision on the measured values and their characteristics should be conducted by the technologist.

2.2 Measuring Instruments

2.2.1 Metrological Characteristics of Measuring Instruments

Measuring instrument—a device used to perform measurements, including those connected to one or more additional devices.

Measuring system—a set of one or more measuring instruments and often other devices, including, if necessary, reagents or power supplies, collected and adapted to obtain information about the measured values within the established intervals for the values of these genera.

Measuring instrument with indicating measuring device—a measuring instrument that provides an output signal that carries information about the value of the measured value. The input signal can be presented in both visual and audio form.

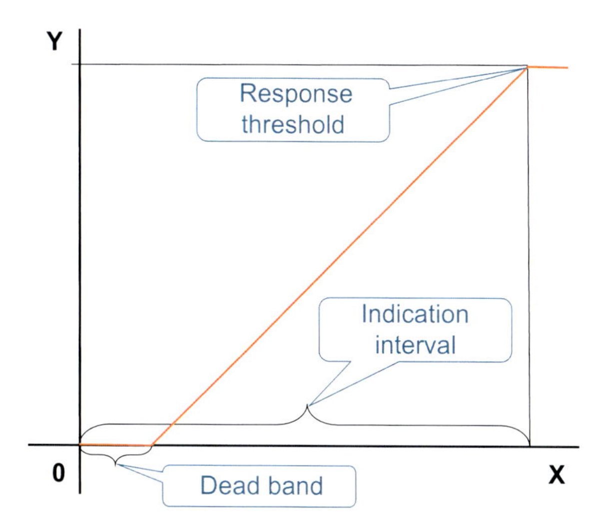

Fig. 2.2 Characteristics of measuring instruments

Such a tool may also record in one form or another the measurement results and / or transmit the measurement results to another device.

All measuring instruments and measuring systems must meet certain metrological characteristics to ensure that the measurements meet the requirements.

Metrological characteristic—a characteristic feature that can affect the measurement results. The metrological characteristics include [2]:

Indication interval—a set of values, limited by the maximum possible readings. This interval is usually determined by the largest and smallest values of this value, for example "from 0 to 82 V" (Fig. 2.2). The nominal range of readings is achieved by a certain adjustment of the controls of the measuring instrument and is limited to rounded or approximate values of the maximum possible readings, for example, "from 0 to 80 V".

Sensitivity of a measuring system—the ratio of changes in the readings of the measuring system to the corresponding change in the value of the measured quantity.

The dead band is the maximum interval within which the value of the measured value can be changed in both directions without causing a noticeable change in the corresponding reading.

Discrimination threshold—the largest change in the value of the measured value, which does not cause a significant change in the corresponding reading.

Resolution—the smallest change in the measured value, which is the cause of a significant change in the corresponding reading.

Step response time is the period between the moment when the input value of a measuring instrument or measuring system undergoes a sudden change between two specific constant values, and the moment when the corresponding reading is set within certain limits around its constant final value.

Accuracy class—a classification characteristic of measuring instruments or measuring systems that meet the established metrological requirements, which are designed to maintain measurement errors or instrumental uncertainties within the established limits under certain operating conditions. The accuracy class is usually denoted by a number or symbol accepted by agreement. The accuracy class is also applied to material measures.

Maximum permissible measurement error—the limit value of measurement error relative to a known reference value, allowed by the specification or regulations for a given measurement, measuring instrument, or measuring system. Usually, when there are two limits, the terms "maximum permissible error" or "permissible error limit" are used.

Measurements are often accompanied by the consumption of energy from the object or the introduction of energy into the object whose parameters are being measured. This energy should be minimal, as it affects the value of the measured quantity.

The given metrological characteristics of measuring instruments can change depending on the value of the measured quantity and the speed or frequency of its change. This requires correction of the results obtained.

2.2.2 Measurement Standard, Reference and Working Measuring Instruments

The accuracy of measurements is largely determined by the metrological characteristics of the measuring instruments used.

To ensure proper measurement accuracy, all measuring instruments must be calibrated. National legislation may require a number of categories of instruments to be officially certified by the relevant authorized state metrological services for their intended use.

The conclusion regarding the suitability of measuring instruments is made based on the results of control of their metrological characteristics for compliance with the requirements of regulatory and technical documentation.

Hierarchically, measuring instruments are divided into standards and working.

Standard (measurement standard, etalon)—the implementation of the definition of a given value with a set value and the associated measurement uncertainty, which is used as a basis for comparison. For example, a system of measuring instruments for reproducing a meter through the speed of light propagation in a vacuum, approved as the state standard of the meter.

Metrological standards are divided into several groups.

International standard (international measurement standard)—a standard that is recognized by all states that have signed an international agreement and is intended for global application. An international prototype of a kilogram can serve as a textbook example.

National standard (national measurement standard)—a standard recognized by national authorities for use in the state or in the economic system as a basis for assigning values to other standards for this type of value.

The status of national standards is granted (usually by the government) to both primary and secondary standards if there is no corresponding primary standard in the country.

Primary measurement standard—a standard based on the use of a primary reference measurement technique or created as an artifact selected by consent.

Secondary measurement standard is a standard that is calibrated according to the primary standard for a value of the same kind. Used to reduce the load on the primary standard.

Source reference or **reference standard (reference measurement standard)**—a standard designed to calibrate other standards for values of this kind in this organization or in this place.

A **working measurement standard** is a standard used for the daily calibration or verification of measuring instruments or measuring systems. Working standards are used in metrological centers, calibration and calibration laboratories. The working standard is usually calibrated according to the original standard.

Working measuring instruments may not be used to verify other measuring instruments if they are even more accurate than existing working standards, as they have not been officially approved. However, working standards may not be used as working measuring instruments to perform practical measurements.

A **reference material** is a material that is sufficiently homogeneous and stable with respect to certain properties that have been established to be suitable for its intended use in the measurement or study of qualitative properties.

The study of a qualitative property provides the value of this property and the associated uncertainty. This uncertainty is not the uncertainty of the measurements. Standard samples with or without assigned values can be used to control the accuracy of measurements, whereas only standard samples with assigned values can be used to calibrate or control the accuracy of measurements. The concept of a standard sample includes materials that embody both size and quality characteristics.

2.2.3 Analog Measurement Instruments

Analog measuring instruments are designed to measure continuous physical quantities and contain several typical devices (Fig. 2.3).

The conversion device contains a **sensor**—an element of the measuring system, which is directly affected by the phenomenon, body or substance that is the carrier of the quantity to be measured. Depending on the nature of the measured quantity, the parameters of the measured quantity to be determined, and the method of their measurement, the conversion device may contain several sensors, scale, functional and other types of measuring transducers.

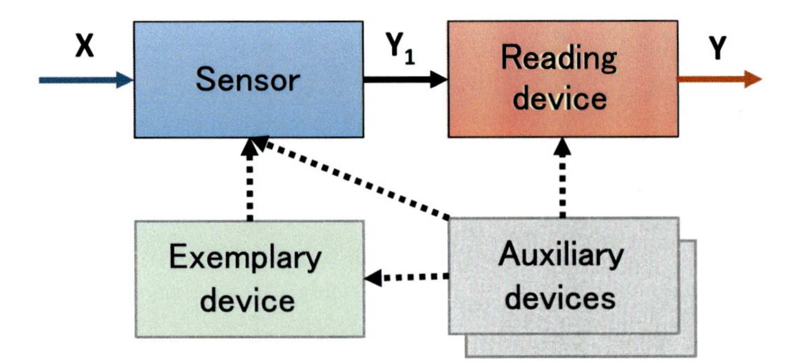

Fig. 2.3 Structure of analog measuring instrument

The reading device is designed to convert the output signal of the conversion device Y_1 into a form suitable for reading the values of the measured value.

An exemplary device may be part of the analog measuring instrument if it is necessary to calibrate the latter during operation.

Auxiliary devices, like the sample device, are not mandatory elements of analog measuring instruments and do not participate in the measurement process. These include power supplies, acoustic alarms, lighting devices, overload fuses, etc.

The components of analog measuring instruments are inertial elements and, accordingly, the readings of the measuring instrument depend on the frequency of change of the measured value. Therefore, among the metrological characteristics of the measuring instrument indicate the frequency range in which these characteristics are provided.

2.2.4 Digital Measuring Instrument

Digital measuring instruments are designed to measure physical quantities and contain several typical devices (Fig. 2.4).

Like analog measuring instruments, digital ones contain an input analog converter. Depending on the nature of the measuring quantity and the method of measuring,

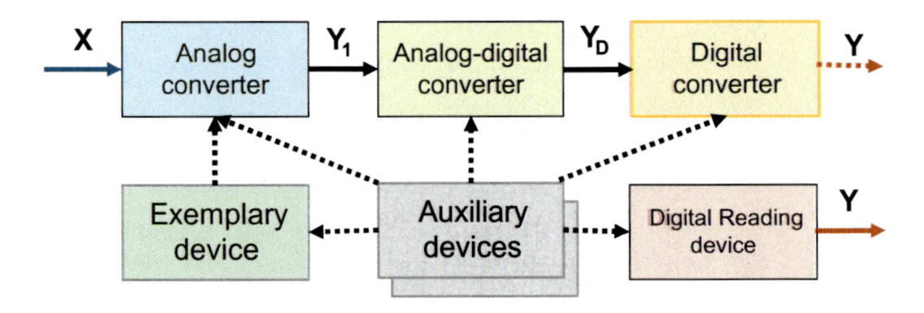

Fig. 2.4 Structure of digital measuring instrument

the conversion device may contain one or more sensors, scale, functional and other types of measuring transducers.

Digital measuring instruments include an analog-to-digital converter (ADC)—a device that converts the input analog (continuous) signal into a discrete code (digital signal). The code can be converted according to the method of measurement and the parameters of the measured value to be determined, a digital converter, often built on a microprocessor. The digital converter can also provide the transfer of measurement results to other devices according to certain data transmission protocols.

The digital reading device provides the formation of a visual signal of the measurement information, which is presented in the form of numbers or symbols.

Auxiliary devices, like the sample device, are not mandatory elements of digital measuring instruments and do not participate in the measurement process. These include power supplies, acoustic alarms, lighting devices, overload fuses, etc.

The components of digital measuring instruments are inertial elements and, accordingly, the readings of the measuring instrument depend on the frequency of change of the measured value. However, in addition to the inertia of the elements of the tool. Of great importance is the frequency of analog-to-digital conversion in the measuring instrument. According to the Whittaker-Nyquist-Kotelnikov-Shannon sampling theorem, to recover a signal from its lossless samples, it is necessary that the sampling frequency be at least twice the maximum frequency of the primary continuous signal.

$$F_{ADC} = 2F_{max}. \tag{2.5}$$

Therefore, among the metrological characteristics of the measuring instrument indicate the frequency range in which these characteristics are provided considering the frequency of analog-to-digital conversion.

2.3 Measuring Methods

2.3.1 Electrical Parameters

Most welding processes involve the use of electricity as a source of heat or as a source of energy for ancillary equipment. Therefore, the measurement of current and voltage is an integral part of any technical requirements for welding procedure specifications (WPS).

Measurement of current and voltage is conducted by measuring instruments called ammeters and voltmeters, respectively (Fig. 2.5).

To measure the current, the ammeter A is connected in series to an electrical circuit. Since the measuring instrument must have a minimal effect on the object of measurement, the electrical resistance of the ammeter is minimal and, ideally, close to zero.

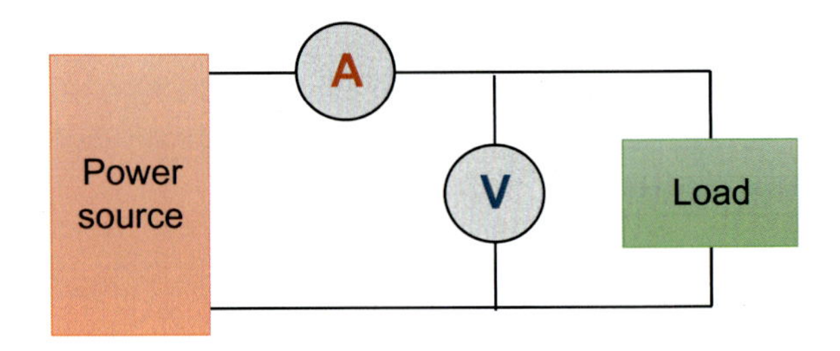

Fig. 2.5 Measuring current and voltage

The electrical voltage is measured between two points in an electrical circuit. For minimal impact on the object to be measured, the electrical resistance of the voltmeter should be very significant, ideally going to infinity.

Both direct and alternating current are used in welding and related processes. It should be noted that the term "direct current" in welding often means only that the current is unipolar. The current amplitude can vary considerably, as occurs when welding with a fusible electrode in a protective gas environment.

The range of currents used in various welding methods is from units of milliamperes (electron beam welding) to tens of kiloamperes (resistance welding). Accordingly, different types of current and voltage measuring instruments are used, which are based on different physical effects.

Expansion of the DC voltage measurement range is carried out by using additional resistance, by including a resistor R_{ad} in the measuring circuit (Fig. 2.6).

When using additional resistance, the measured voltage value is converted from the readings of the U_V voltmeter by the formula:

$$U = U_V \cdot \left(1 + \frac{R_a}{R_V}\right). \tag{2.6}$$

Additional resistors can be built into the voltmeter or be a separate device. The accuracy class of the additional resistor must be higher than the accuracy class of the voltmeter.

Expansion of the range of DC measurement is carried out by using a shunt R_s in the measurement circuit (Fig. 2.7).

The current in the electrical circuit is determined by the voltage drop across the shunt. Since the shunt must have low electrical resistance, the voltage drop across the shunt is measured with a millivoltmeter.

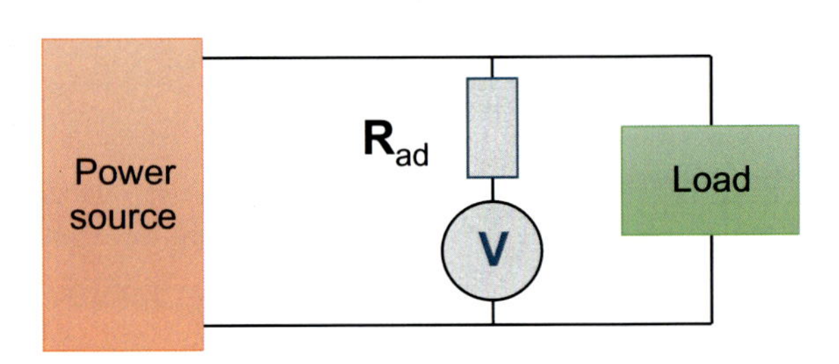

Fig. 2.6 Voltage expansion measuring diagram

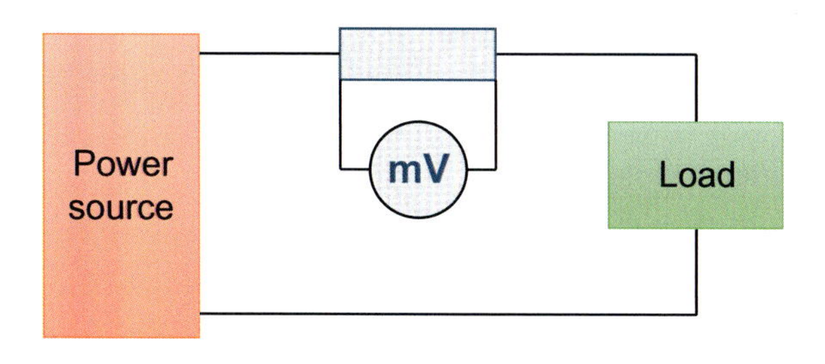

Fig. 2.7 Scheme of expansion of DC measuring

$$I = \frac{U_{mV}}{R_{\text{III}}}. \tag{2.7}$$

Structurally, the shunt is a flat resistor made of an alloy having a stable resistance in the operating temperature range with the attachment of power wires and potential clamps (Fig. 2.8). The millivoltmeter is connected to the shunt by calibrated wires with guaranteed resistance.

Calibrated external shunts are available at different rated currents (from 1.5 to 10,000 A). The voltage drops across such shunts during the rated current are 45, 60, 75, 100 or 300 mV, depending on the brand.

With this method of measuring the ammeter is essentially a set consisting of a shunt and a millivoltmeter.

In welding equipment, means for measuring shock and stress are often used, in which the reading device is an arrow moving relative to the scale with divisions. The most common are means of magnetoelectric, electromagnetic and electrodynamic systems.

The magnetoelectric system is suitable for measuring only direct current or voltage, electromagnetic and electrodynamic systems can be used for both alternating and direct current.

Indicator instruments for measuring alternating current of electromagnetic and electrodynamic systems have a non-uniform scale, which is graduated in the current (rms) values of alternating sinusoidal current. Deviation of the measured value from the sinusoidal law of change for such measuring instruments leads to a significant loss of accuracy—up to 20% when welding under flux with alternating current.

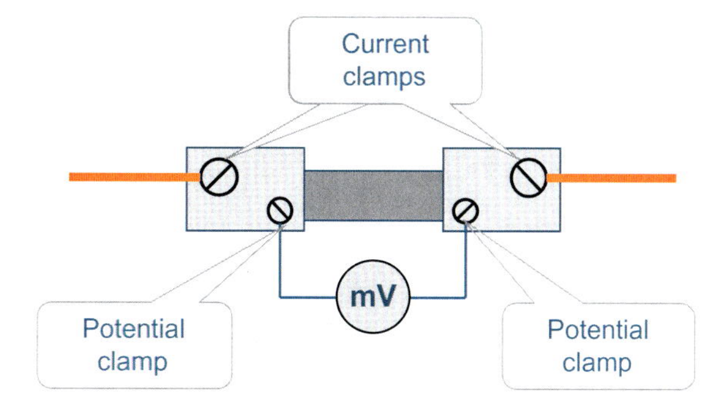

Fig. 2.8 External current shunt

Measuring transformers

Measuring transformers are used to expand the limits of **AC measurement**. In addition to expanding the range, measuring transformers allow galvanic separation of the measuring circuit and the power circuit.

Measuring transformers consist of two isolated windings placed on a ferromagnetic core (Fig. 2.9). The primary winding has the number of turns W1, the secondary—W2. The terminals of the primary winding are connected to the measuring circuit, and an ammeter is connected to the terminals of the secondary winding.

The primary winding of the current transformer W1 contains from one to ten turns and can be either part of the transformer or made directly by a welding cable that moves into the window of the core. The secondary winding contains a large number of turns, i.e. the measuring current transformer is a step-up transformer. Because of this, the secondary circuit of current transformers must be earthed for safe operation. When the ammeter is switched off, the secondary winding must be short-circuited for safety reasons. In addition, this mode is emergency for the current transformer due to a sharp increase in losses in the ferromagnetic medium and, consequently, its strong heating.

The measured values are determined by the readings of the devices multiplied by the corresponding transformation coefficients. The rated current of the secondary winding is usually 5A.

A variant of the measuring current transformer is the Rogovsky belt (Fig. 2.10). A distinctive feature of the Rogovsky belt is the absence of a ferromagnetic core. Due to the absence of a ferromagnetic medium, the Rogovsky belt is a linear transducer in the entire range of measured currents.

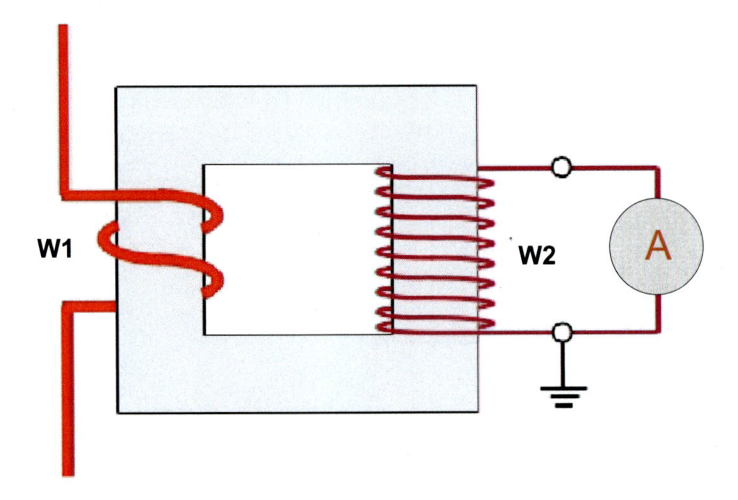

Fig. 2.9 Measuring current transformer

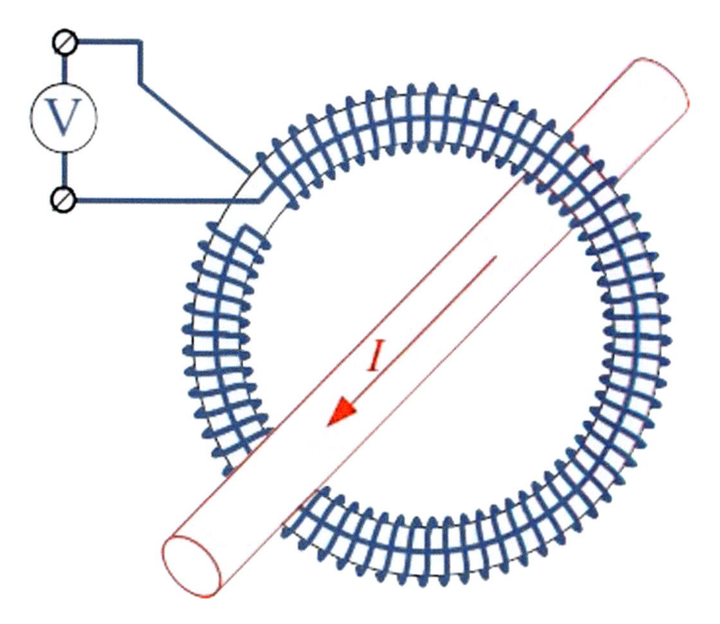

Fig. 2.10 Rogovsky belt

The coil of the Rogovsky belt is wound on a dielectric frame, through the hole of which a current-carrying element passes. Excluding the influence of the coil formed by the winding itself, its end is returned to the beginning as shown in Fig. 2.10. The winding terminals are connected to a voltage meter.

The Rogovsky belt is sensitive to the rate of change of the magnetic field created around the conductive element. passing it. Thus, the DC component of the current cannot be detected by the Rogovsky belt.

In welding, the Rogovsky belt is most often used to measure the current in the secondary circuit of transformers of resistance spot welding machines. Current transformers with a ferromagnetic core are not suitable for this, because the introduction of the ferromagnetic mass into the welding circuit of such a machine greatly reduces the current in it.

The current arising in the secondary winding of any measuring current transformer is proportional to the rate of change of current in the primary winding, i.e.

$$i_2 = f\left(\frac{di_1}{dt}\right) \tag{2.8}$$

When measuring sinusoidal current, this feature of current transformers has almost no effect on accuracy. However, the violation of sinusoidality, which inevitably occurs during arc welding butt welding with thyristor current control errors increase sharply and can exceed 20%. This error can be reduced by applying analog-to-digital conversion followed by computer data integration.

Measuring instrument for electrical parameters based on Hall effect

Unlike current transformers, measuring instruments built on the application of the Hall effect make it possible to measure both alternating and direct current.

The Hall effect is as follows. Place the conductive plate in the form of a parallelepiped perpendicular to the magnetic field with a squint of H (Fig. 2.11). Connect two opposite faces (points 1 and 2) to the current source E and pass a direct current I through the plate. Then, due to the interaction of charge carriers with the magnetic field, between the other two faces (points 3 and 4) there is a potential difference U.

In an ammeter built on this effect, the Hall element is placed in the magnetic field around a conductive element (Fig. 2.12). The voltage generated at the respective faces of the Hall element is amplified to create a compensating winding of the magnetic flux, the opposite of the flux created by the measured current. This allows to expand the measuring range and ensure high accuracy over a wide temperature range.

Visualization of the measured value is carried out by analog or digital measuring instruments connected according to Fig. 2.12.

The voltmeter constructed with use of Hall effect works similarly (Fig. 2.13).

In a voltage measuring instrument, a magnetic flux proportional to the voltage is generated by a winding that is connected to the measured circuit through a resistor.

Bridge measurement schemes

Bridge circuits are used to measure the parameters of electrical elements by the method of comparison. Such schemes are characterized by high sensitivity, accuracy, a wide range of measured values of the parameters of the elements.

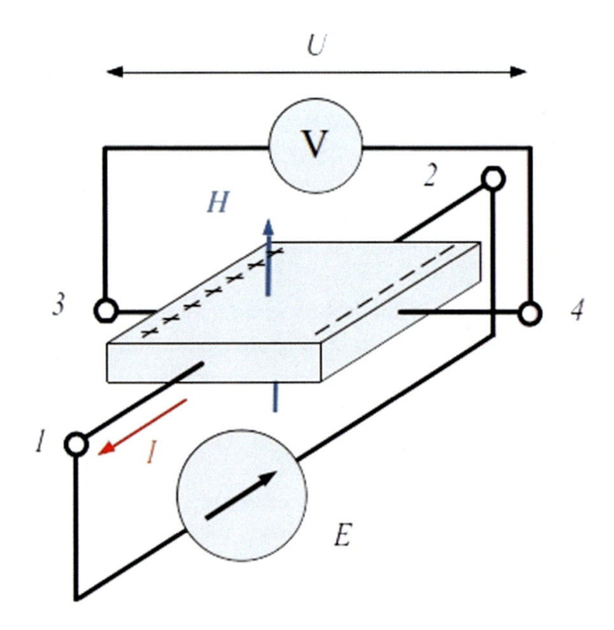

Fig. 2.11 Hall effect

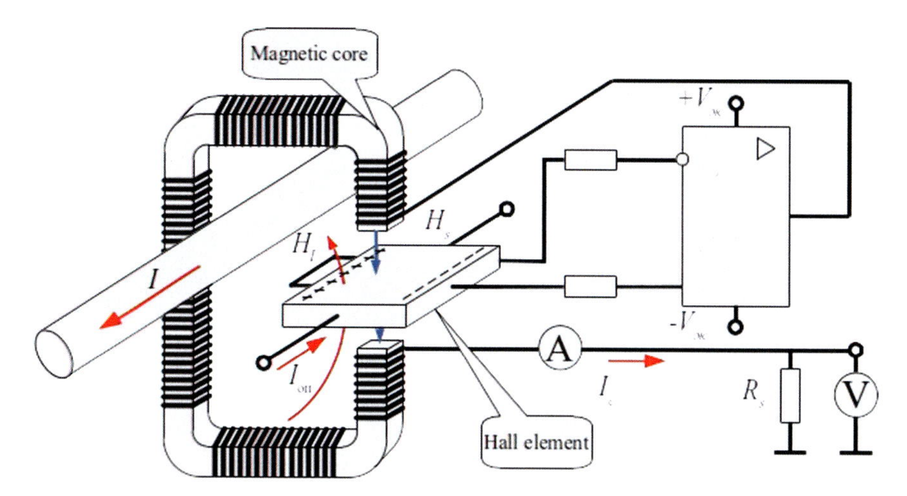

Fig. 2.12 Amperemeter based on Hall effect

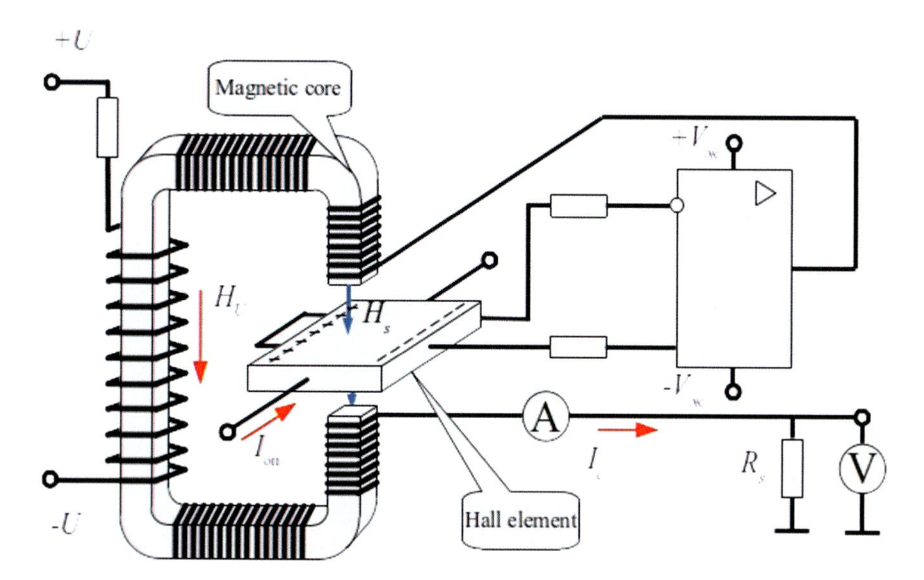

Fig. 2.13 Voltmeter based on Hall effect

The standard bridge circuit, named after the inventor of the Wheatstone Bridge, consists of four resistors (Fig. 2.14).

Each of the four resistors in the bridge circuit is called a lever. Resistors R_1 and R_2 are called the arms of the relationship. Resistance of resistor R_x is unknown. R_3 is commonly referred to as a bridge rheostat.

If the potential difference between point 1 and point 2 is zero, the millivoltmeter will show zero that is called "balance".

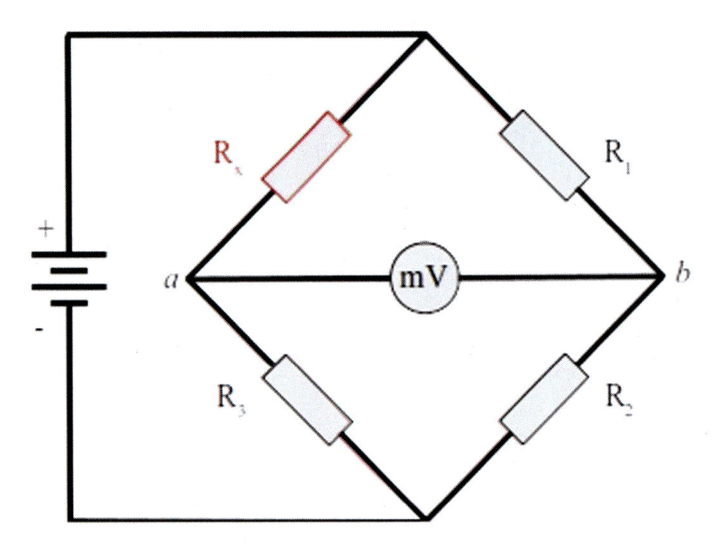

Fig. 2.14 Wheatstone Bridge

The balance of the bridge depends only on the ratio of the resistors' resistance and does not depend on the voltage. The balance of the bridge requires the ratio to be met:

$$\frac{R_x}{R_3} = \frac{R_1}{R_2}. \tag{2.9}$$

Each of these three resistors can be replaced by a resistance of another value or their values can be adjusted so that the bridge is balanced, and when this happens the value of the resistance of the unknown resistor can be determined from the ratio of known resistances. To do this, the resistances of three resistors must be known.

2.3.2 Pressure Measurement

In technological processes of welding (electron beam, diffusion, resistance spot, etc.) pressure is one of the parameters of the process. In welding technologies, excess pressure and vacuum are most often controlled.

Pressure gauges are called manometers. The primary pressure transducers of the respective measuring instruments most often convert the pressure into displacement.

Spring or deformation primary transducers are most often used in welding technology. The change in pressure leads to a certain deformation of the elastic element. Bourdon tube, elastic membrane, and bellows can be used as an elastic element.

The manometer with a Bourdon tube contains an oval tube curved in an arc of a circle (usually at an angle of about 250°), filled with liquid (Fig. 2.15). Under the

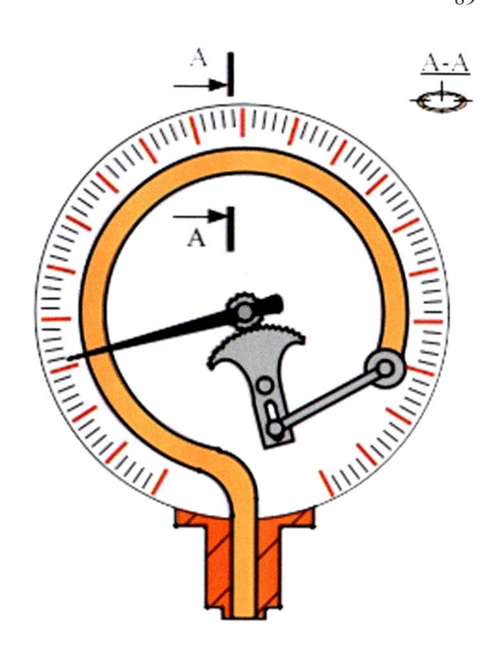

Fig. 2.15 Manometer with a Bourdon tube

influence of the applied pressure the pipe straightens. The needle of the manometer through the system of levers and gear sectors moves in proportion to the applied pressure. The measuring range of such manometers with a Bourdon tube is from 10 to 1000 MPa.

A manometer with an elastic membrane (Fig. 2.16) converts its deformation under the pressure into displacement. The deformation of the membrane is directly proportional to the applied pressure. The needle of the manometer moves through a system of levers and gear sectors. Measured pressure range: from 0.0016 to 4 MPa.

The manometer with a bellows as well as the previous converts deformation in movement. Measured pressure range: from 0.1 to 60 MPa.

To obtain an electrical output signal proportional to the pressure, it is necessary to mechanically move the sensitive element of the manometer into an electrical signal. However, you can get the desired signal directly by using strain gauges. In such a manometer, the sensitive element is a thin-walled steel cylinder with glued strain gauges (Fig. 2.17). The deformation of the cylinder walls is directly proportional to the pressure and its diameter and inversely proportional to the wall thickness and the modulus of elasticity of the material. The deformation is perceived by the strain gage R_T. Corrective strain gage R_c is designed to compensate for temperature deformations. Changing the resistance of strain gages under the action of medium pressure in the middle of the cylinder leads to a change in voltage in the diagonal of the Wheatstone bridge. After amplification, the signal proportional to the pressure can be transmitted to the automatic control circuit or output to the visualization means. In modern measuring instruments, it is also possible to digitize the signal and transmit it to the digital control system using one of the standard data transmission protocols.

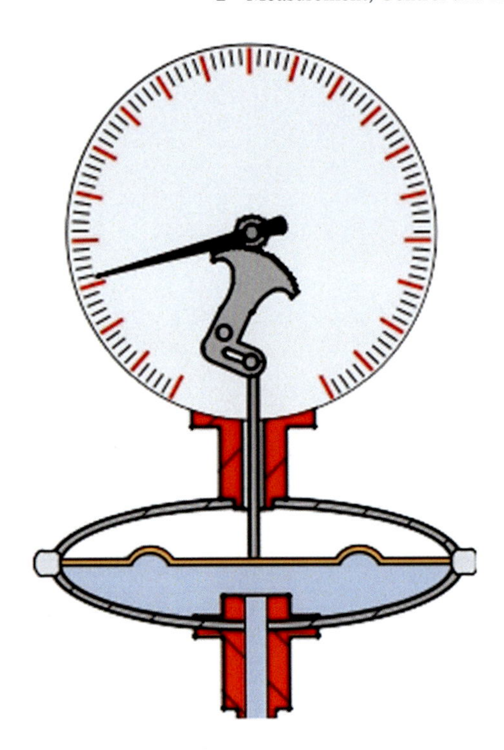

Fig. 2.16 Manometer with an elastic membrane

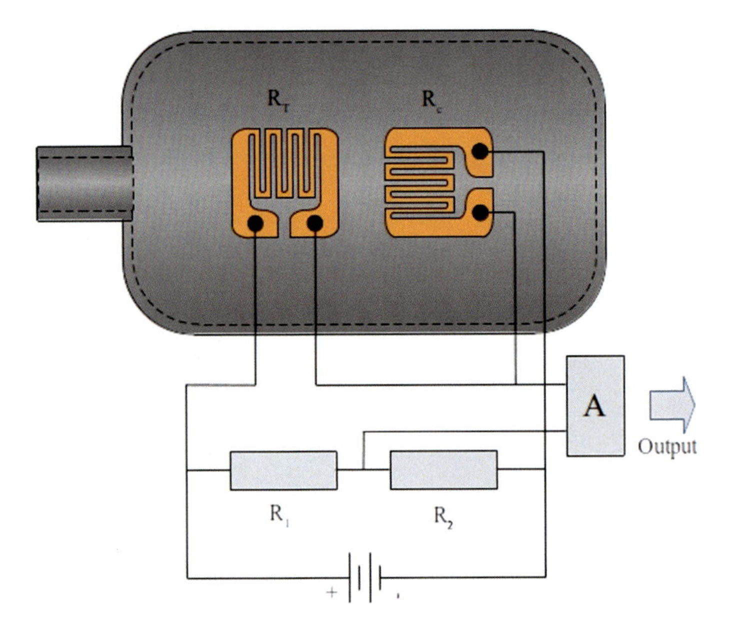

Fig. 2.17 Manometer with strain gauge

2.3.3 Gas Consumption

Welding in shielding gases requires stable local protection of the welding arc zone and the formation of a joint. The protection is carried out by a jet corresponding to the selected technology of welding gas or gas mixture, which is fed into the melting zone through the torch. The quality of jet protection is determined by the design of the torch nozzle, the distance from the nozzle cut to the welding pool and the consumption shielding gas. Proper protection of the arc and the zone of formation of the connection is guaranteed only by laminar gas leakage from the nozzle. In addition to the design of the torch and its correct position relative to the product to be welded, the quality of protection is ensured by the proper gas consumption. It should be noted that exceeding a certain limit of gas consumption, which is determined by the design of the torch, leads to turbulence of the gas flow and significant deterioration of protection.

The most common way to provide gas welding stations is cylinder supply. The gas in the cylinders is compressed or liquefied. From this follows the need to stabilize and regulate the pressure of the gas entering the burner. The standard means for this is a gas reducer (Fig. 2.18).

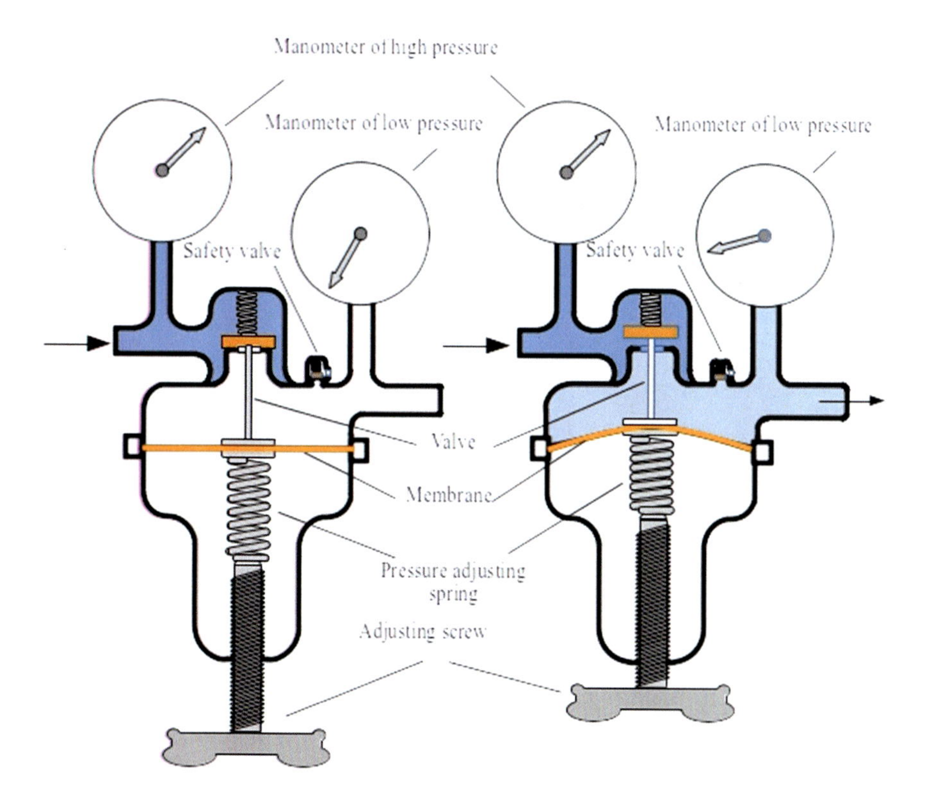

Fig. 2.18 Gas reducer operating scheme

The reducer is connected to a cylinder in which the gas is under high pressure. The gas enters the high-pressure chamber, which can be monitored with a suitable manometer. The high-pressure chamber is closed by a valve that is pressed against its seat by the spring and gas pressure. When tightening the adjusting screw on the valve, the force of compression of the spring acts and it opens. Gas from the high-pressure chamber enters the low-pressure chamber, which is separated from the environment by a flexible membrane. The increase in pressure in the low-pressure chamber leads to bending of the membrane, compression of the control spring and closing of the valve. Thus, a pressure proportional to the degree of twisting of the adjusting screw is set in the low-pressure chamber. This pressure does not depend on the gas pressure in the cylinder. When the pressure in the low-pressure chamber decreases, the diaphragm returns to its original state, the valve opens and the cycle repeats.

Stable pressure at the outlet of the gearbox provides stable gas flow through the torch. But since different torches have different effective cross-sectional areas of the gas supply system, it is not easy to determine gas consumption.

To measure gas consumption, the reducer is sometimes supplemented with a rotameter (Fig. 2.19). The principle of operation of the rotameter is to balance the float of a certain mass, located in a conical transparent tube, the dynamic pressure of the gas. The losses are determined by the position of the float relative to the scale applied to the tube.

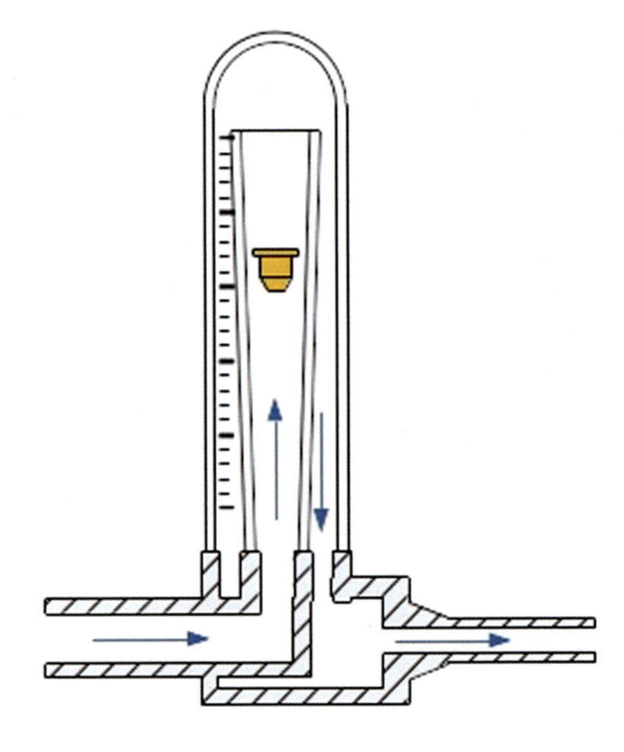

Fig. 2.19 Rotameter principle of operating

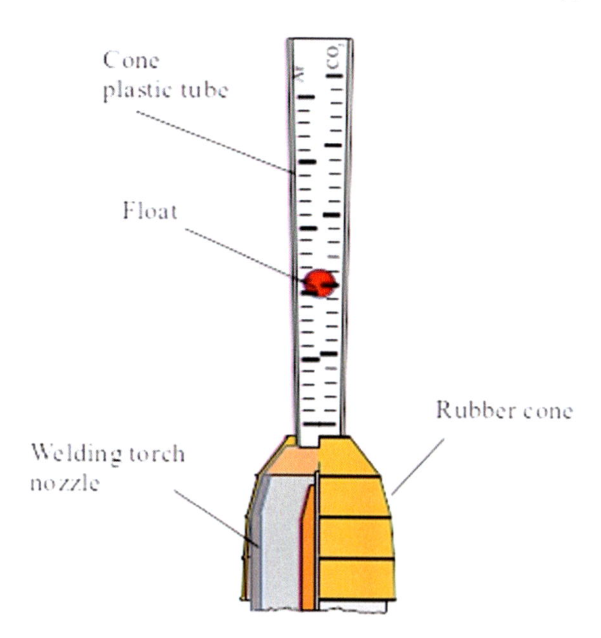

Fig. 2.20 Control rotameter

To control the gas flow directly through the torch nozzle, a rotameter of a slightly different design is used (Fig. 2.20).

When using rotameters, they must be placed strictly vertically. The rotameter readings are significantly affected by the temperature and nature of the gas. That is why calibration curves are added to the rotameter kit. Sometimes several scales for different gases are applied to the rotameter (Fig. 2.20).

2.3.4 Welding Speed

The welding speed can be determined by the length of the obtained weld and the time spent on its execution. However, this method gives only the average speed and does not allow operational control and correction if necessary. Tachogenerators and encoders are used for operative control of welding speed.

The tachogenerator is an electric measuring generator that converts the speed of the shaft into the corresponding frequency of the electrical signal. Tachogenerators can be DC and AC. DC welding tachogenerators are most often used in welding installations as they have the smallest dimensions and the output signal (DC voltage) is easier to visualize and process (Fig. 2.21).

Tachogenerators do not work well in the range of low angular speeds of rotation of the shaft and therefore, most often, are connected directly to the shafts of the executive motors of welding equipment. Another disadvantage is the presence of friction pairs in the structure and, accordingly, the need for periodic maintenance.

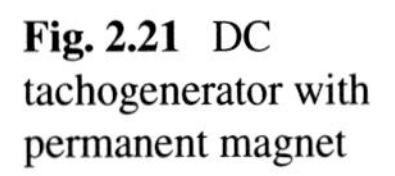

Fig. 2.21 DC tachogenerator with permanent magnet

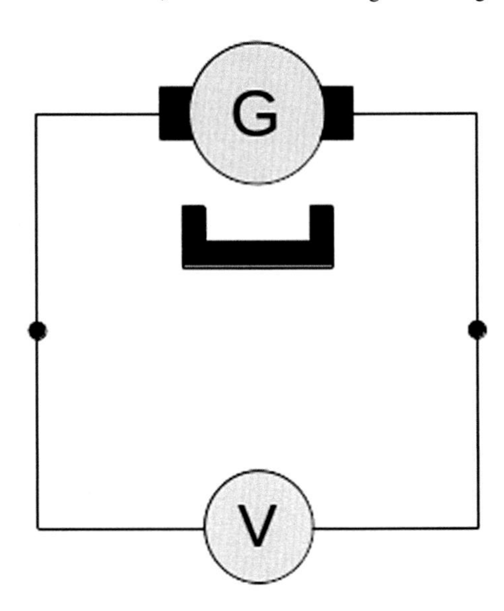

Expanding the range and increasing the accuracy of determining the speed of movement of the working bodies of the welding system is possible with the use of encoders—devices for converting angular or linear movements into digital code. This, however, requires complicating the design of equipment and the use of microprocessor control systems.

Encoders are divided into two types:

- absolute encoders—those providing information about the position relative to a certain point in space.
- incremental encoders—those providing information about the movement of the elementary step relative to the previous position.

Encoders use different physical principles to obtain information about the elementary movement. According to these principles, encoders are divided into:

- electromechanical,
- resistive,
- optical,
- magnetic,
- electromagnetic,
- capacitive,
- other.

Consider the operation of the optical angular absolute encoder (Fig. 2.22). The principle of operation of such an encoder is to scan the luminous flux and convert it into a sequence of digital signals.

When the encoder shaft rotates, the code disk also rotates. The disk contains a set of transparent and non-transparent labels, which determine its current position.

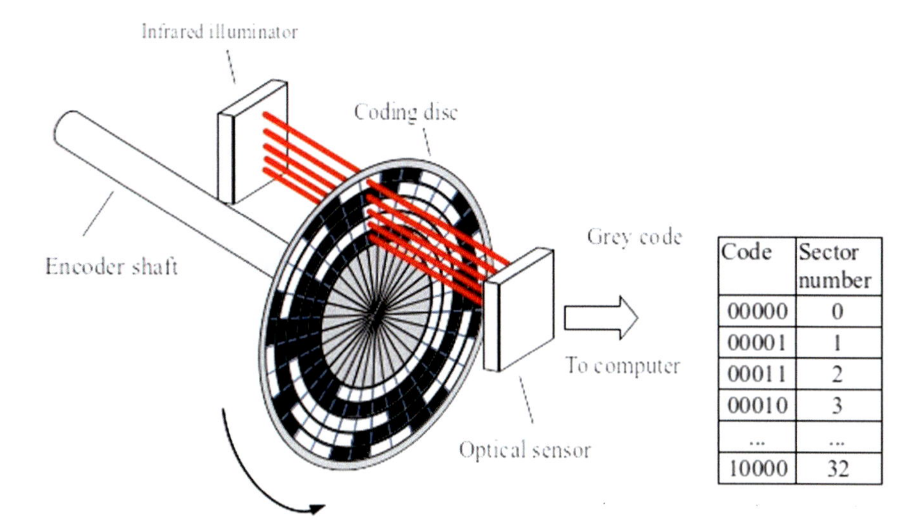

Code	Sector number
00000	0
00001	1
00011	2
00010	3
...	...
10000	32

Fig. 2.22 Absolute optical encoder operating principle

The light from the infrared illuminator enters through an encoding disk on an optical sensor, which generates a digital code corresponding to the position of the disk. To increase the noise immunity, the coding disk is marked so that the optical sensor receives a Gray code at the input. The digital code is transmitted to the controller, which can determine both the position of the working body of the welding plant and the speed of its movement.

Absolute encoders can be multi-rotating due to a suitable mechanical system.

In incremental encoders coding disks with labels that are evenly distributed on the disk are used (Fig. 2.23). Additionally, the encoding disk may have start marks to facilitate the determination of the start of the reference. The optical sensor generates two digital signals that are shifted by half a period relative to each other to determine the direction of rotation of the shaft. Thus, when rotating the shaft in one direction, the front of signal A corresponds to a high level of signal B. When changing the direction of rotation of the edge of signal A will correspond to a low level of signal B.

Linear encoders contain a read head and a guide with a tape on which the code is applied. The principles of signal formation about the position of the head relative to the coding tape are like the principles used in angular encoders. The absence of mechanical contact protects the encoder components from wear.

Linear encoders are widely used in numerically controlled machines, in plasma and laser cutting systems.

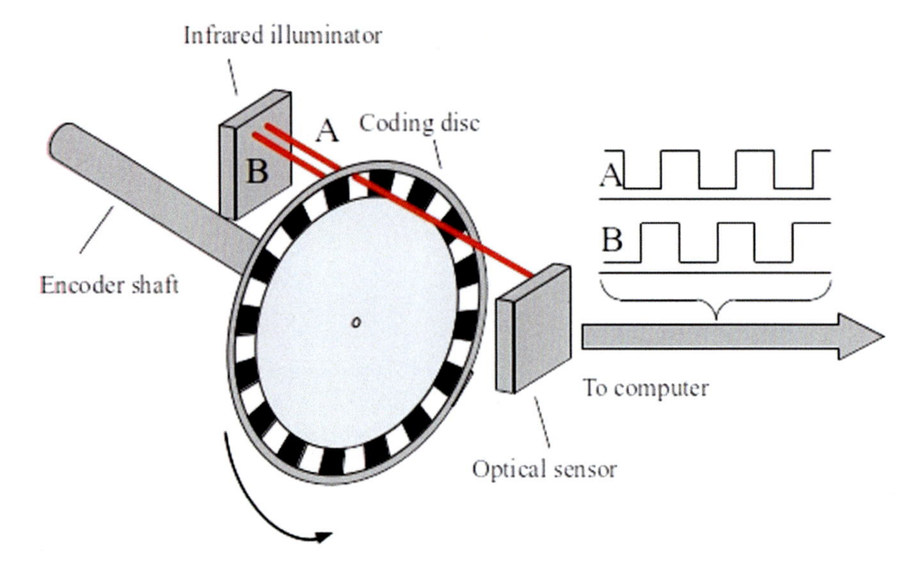

Fig. 2.23 Increment optical encoder operating principle

2.3.5 Temperature (ISO 13916), Humidity, Wind

Several welding technologies require provision and control of a certain temperature regime in the area of joint formation. Thus, the ISO 13916 [3] standard sets out the requirements for the operations of measuring the preheating temperature, the transition temperature and the preheating temperature at the endurance that occurs when the welding process is interrupted, for fusion welding technology.

ISO 13916 [3] defines the requirements for location, time of measurement and measuring instruments.

Temperature measurements are performed on the surface of the so-called structure from the location of the heat source in accordance with WPS. The measuring point must be at a distance A:

$$A = 4 \cdot t. \tag{2.10}$$

where t is the thickness of the weld metal (Fig. 2.24). However, the distance A should not exceed 50 mm for metal thicknesses up to 50 mm.

In cases where the thickness of the base metal exceeds 50 mm, distance A must be at least 75 mm. If possible, the temperature should be measured from the side of the structure opposite to the heating surface. If it is not possible to measure the temperature from the opposite to heating side, the temperature is measured from the heating side after the heating has stopped and the temperatures were equalized. The time of temperature equalization is determined at the rate of 2 min for every 25 mm of thickness of the base metal.

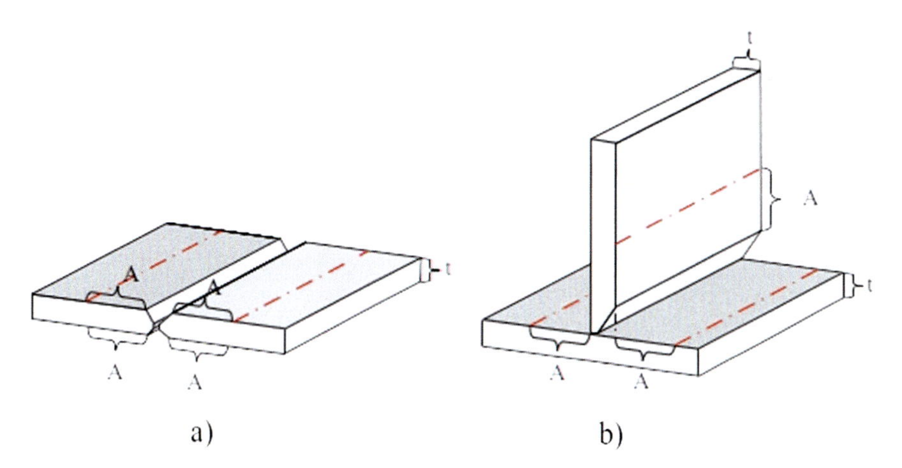

Fig. 2.24 Places of temperature measuring

If constant heating and temperature measurement from the side opposite to the heating surface is impossible during welding due to design features, the temperature is measured from the heating side of the base metal immediately after preparation for welding.

Of course, sometimes the design features of the product can prevent the implementation of these recommendations. In such cases, temperature measurement can be carried out in other places of the welded structure, which are stipulated by the relevant normative document.

In multi-pass welding, the temperature between the passes is measured on the weld metal or in the area of the base metal directly adjacent to the weld.

The temperature measurement procedure must determine the appropriate measuring instruments.

Heat-sensitive materials

Thermoindicators—substances that change their appearance at a certain temperature—can be used to measure temperature. Thermoindicators, widely used in the manufacture of welded structures, are divided into two groups:

1. compositions that change color at a certain critical temperature,
2. compositions that melt at a certain temperature.

Thermal indicators are also divided into reversible and non-reversible. Reversible return to their original state when the temperature decreases.

The first type includes special coatings containing heat-sensitive pigments. Thermal indicators of this type can be both reversible and non-reversible.

Such compositions are available in the form of thermal signal paints that dry quickly. Dry paint changes its color depending on the surface temperature on which it is applied. Paints are available at different nominal temperatures in the range from 30 to 1280 °C. The color transition of such paints is ±2 °C.

The second type of thermal indicators includes pencils, varnishes, tablets and others that contain components, which become transparent when melted. Thermal indicators of this type are irreversible.

Thermoindicator pencils are designed to determine the surface temperature of solids. The basis of the thermal indicator pencil is a rod made of lacquer-polymer composite that melts at a certain (nominal) temperature. At temperatures below the nominal, the label is dry and crumbly. At a surface temperature corresponding to the nominal, the label melts and turns into a glossy smear. The melting of the label indicates the achievement of the nominal temperature, and the change in color is not an informative sign. It is not possible to reuse the applied label, because after cooling the label does not change its appearance.

Thermoindicator pencils provide measurement accuracy $\pm 1\%$ in the range of 40–200 °C.

Contact thermometers

The principle of operation of contact thermometers is based on the change of physical characteristics of materials under the influence of heat from the object under study. From the very name "contact thermometer" it follows that for the operation of such thermometers it is necessary to ensure reliable thermal contact of the sensitive element of the measuring instrument and the object whose temperature is to be measured.

Most often, contact thermometers use the following physical characteristics:

- thermal expansion of liquids or solids and related phenomena,
- change of electrical resistance,
- thermoelectric phenomena.

Thermometers based on expansion of liquids

An example of a contact thermometer that uses liquid expansion when heated is an alcohol thermometer. This thermometer contains a transparent glass tank with a capillary tube soldered to it. The tank is filled with tinted alcohol. The measuring instrument also contains a temperature scale on which the values are read. In welding it is used to control the working environment and storage conditions of materials and equipment. Measuring range -80 to $+50$ °C.

Bimetallic thermometers

The phenomenon of thermal expansion to measure temperature is also used in so-called bimetallic thermometers. The thermobimetal used to measure temperature is a fixed two-layer plate or tape. Thermobimetal layers are made of metals or alloys with different coefficients of thermal expansion.

The main element of the design of the bimetallic thermometer is a piece of bimetallic tape, twisted, as a rule, in a spiral to reduce the dimensions of the measuring instrument (Fig. 2.25). One end of the tape is fixed in the housing of the device. The

Fig. 2.25 Bimetallic thermometer

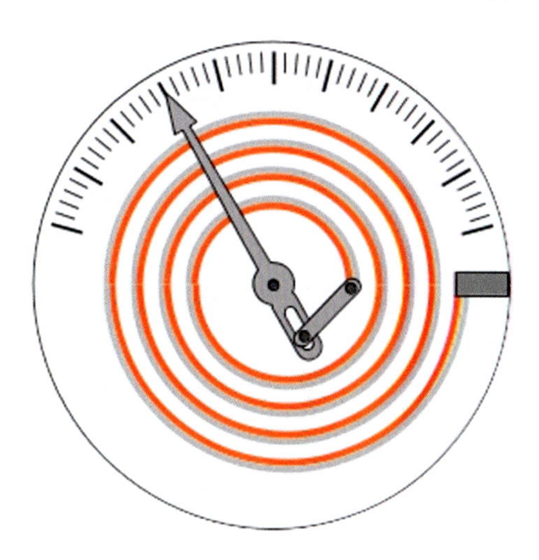

other end of the tape is connected to the arrow of the measuring instrument through a system of levers. When heated, the thermobimetal bends towards the layer with a lower coefficient of thermal expansion and the free end moves, which leads to the rotation of the arrow of the reading device.

The thermobimetallic spiral can be both flat and spatial, depending on the needs of measurement.

Resistive thermometers

The principle of operation of resistance thermometers is the dependence of the electrical resistance of metals, alloys or semiconductor materials on temperature. A typical resistance thermometer is a resistor—a certain number of turns of wire wound on a ceramic core (Fig. 2.26). The wire is made of copper, nickel or, for accurate thermometers, platinum. The resistor is placed in a steel housing filled with electrical sand for electrical insulation.

When the sensitive element is heated, its active resistance changes according to the temperature acquired by it. This resistance is measured using a Wheatstone bridge (Fig. 2.27). A three-wire circuit is used to compensate for errors associated with the resistance of the connecting cables.

The scale of the millivoltmeter can be calibrated in degrees.

Platinum film resistance thermometers are used to measure temperatures up to 660 °C.

Thermoelectrical thermometers

The thermoelectric thermometer actually contains a thermocouple (thermoelectric temperature converter) and a voltage measuring device. The thermocouple is two wires of different metals, connected to each other (Fig. 2.28), usually by welding. When the point of contact between the free and cold ends of the wires is heated, a potential difference appears (Seebeck effect). This potential difference depends on

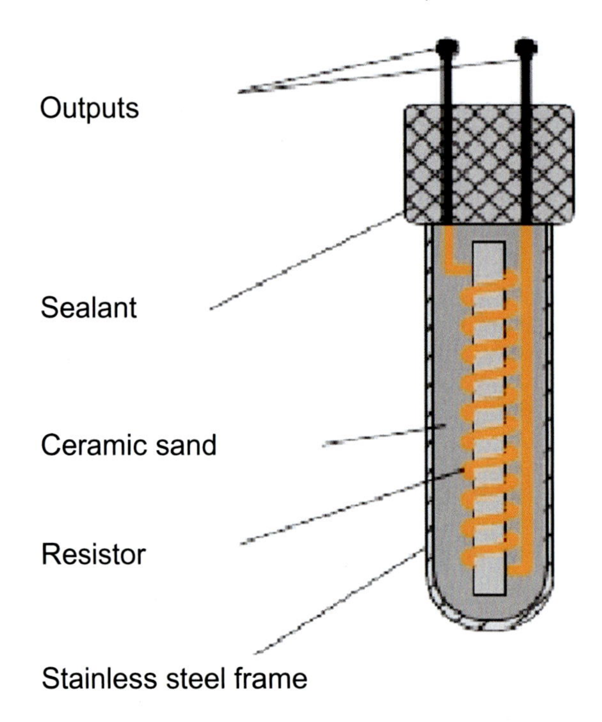

Fig. 2.26 Sensor of resistance thermometer

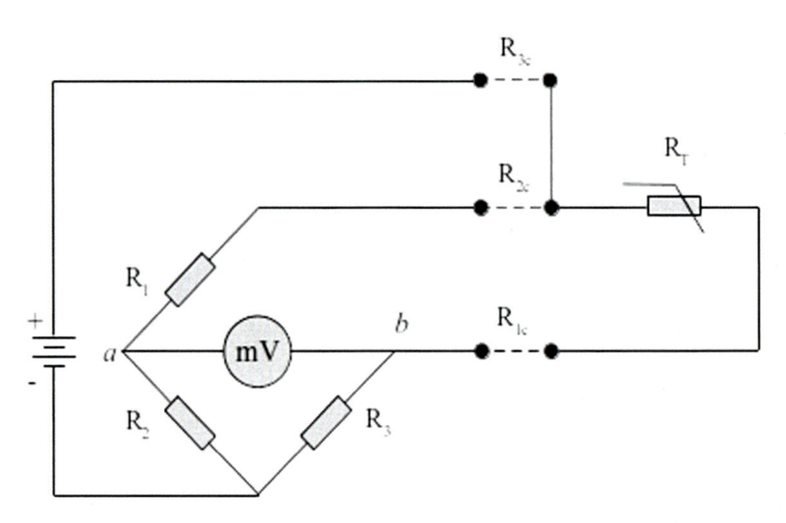

Fig. 2.27 Bridge three-wire connection scheme of the resistance thermometer

the material of the wires and the temperature difference between their cold and hot ends.

The thermoelectric thermometer shown in Fig. 2.28 measures the temperature difference between T_1 and T_2. To determine the actual temperature, the millivoltmeter readings must be corrected accordingly. It should also be noted that the thermoelectric effect is nonlinear.

The measurement of the measuring instrument with respect to the ambient temperature can be adjusted automatically. For this purpose, the bridge scheme is used (Fig. 2.29).

Resistor R_e is made of copper, and resistors R_1, R_2, R_3—of manganin, an alloy whose resistivity depends little on temperature. Resistor R_e is at ambient temperature T_2. If this temperature is different from the temperature at which the calibration took place, the balance of the bridge is disturbed and a corresponding voltage appears between points a and b, which is added to the electromotive force of the thermocouple. The sensitivity of the circuit is adjusted by the resistor R_k.

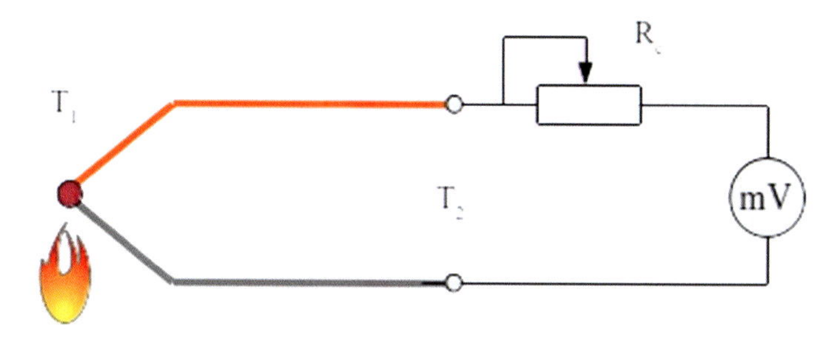

Fig. 2.28 Thermoelectrical thermometer

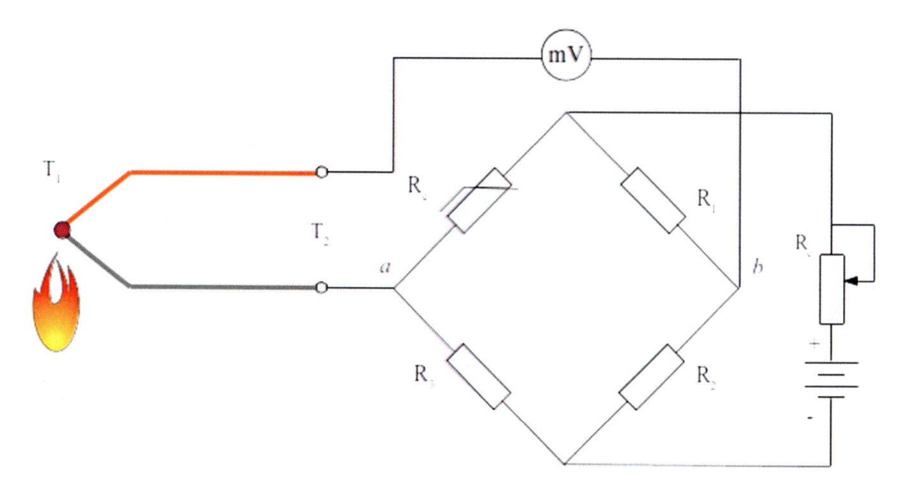

Fig. 2.29 Thermoelectrical thermometer with automatic adjustment

The range of measured temperatures can be up to 1600 °C when suitable materials for thermocouples are used.

Requirements for thermocouples are regulated by international standards:

- IEC 60584-1:2013 Thermocouples—Part 1: EMF specifications and tolerances [4]
- IEC 60584-3:2021 Thermocouples—Part 3: Extension and compensating cables—Tolerances and identification system [5].

Contactless thermometers

Means for temperature measuring by non-contact methods are called pyrometers. Measurements are made by directing the optical system of the measuring instrument at the object. The measurement results are displayed in analog or numerical form. Modern pyrometers are equipped with systems for transmitting measurement results directly to the computer or through the network.

The main technical characteristics of pyrometers, which determine the scope of their application, are the operating temperature range, optical resolution, adjustment of the degree of blackness of the object.

The operating temperature range of the pyrometer is determined by the sensitive element. In general, various non-contact methods allow to measure the temperature from −50 °C to 4000 °C, although, of course, none of the industrial pyrometers cover the entire specified range. The accuracy of measurement by pyrometers averages 2%.

The resolution of a pyrometer is defined as the ratio of the distance from the object to the diameter of the spot from which the radiation from the optical system enters the sensitive element of the measuring instrument. Depending on the pyrometer model, this figure can be in a fairly wide range, from 2:1 to 600:1. For an accurate measurement, the area of the object whose temperature is to be measured must completely fill the measurement spot. Failure to comply with this requirement significantly and unpredictably increases the measurement error. To ensure the required accuracy of the pyrometers, most of them are equipped with aiming systems, including laser.

The radiation coefficient is the ratio of the radiation power of the heat of the investigated surface to the radiation power of an absolutely black body and theoretically it lies in the range from 0 to 1. The radiation coefficient significantly depends on the state of the surface. Thus, stainless steel has a radiation coefficient of 0.85 and the same steel, but polished −0.075. Failure to consider the radiation coefficient leads to significant measurement errors. Therefore, industrial means of non-contact temperature measurement can adjust the radiation coefficient.

The industry uses pyrometers with different principles of temperature determination by non-contact method.

Optical pyrometers are designed to visually assess the temperature of a heated body by comparing its color with the color of a hot reference metal thread (Fig. 2.30).

The filament is in special measuring lamps and is heated by an electric current. When the brightness of the thread becomes equal to the brightness of the object, the thread becomes invisible. This means that its temperature Tn is equal to the temperature of the object T_o. Figure 2.31 shows an image that can be seen in the

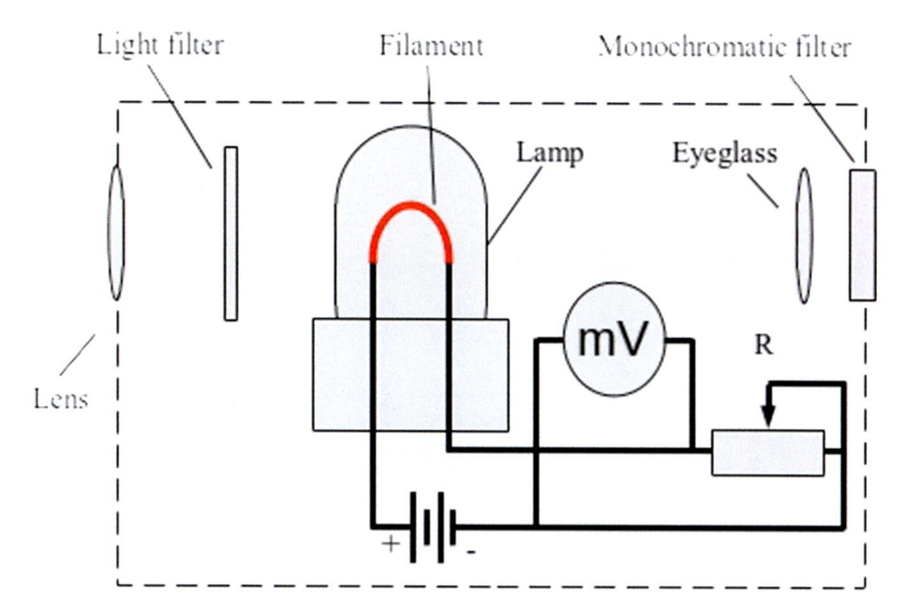

Fig. 2.30 Optical pyrometer

glasses of the optical pyrometer. The color is determined by a monochromatic filter. The temperature of the filament is determined by the current of its incandescence and recorded with a millivoltmeter by the voltage drop across the adjusting potentiometer.

Radiation pyrometers estimate the temperature of the thermal radiation of a heated body in the infrared in a certain range. Pyrometers that estimate radiation in a wide spectral range are called total radiation pyrometers (Fig. 2.32).

The radiation from the heated object is focused by the optical system on a sensitive element—a miniature thermoelectric battery. Thermo EMF is amplified and fixed by a reading device. The pyrometer is equipped with a monochrome filter eyepiece for aiming at the object.

Color or, alternatively, multispectral pyrometers (Fig. 2.33) allow to measure the temperature of the object, based on the results of comparisons of its thermal radiation

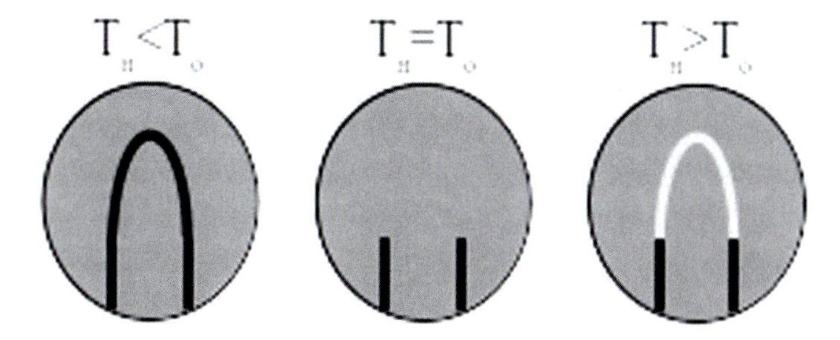

Fig. 2.31 Image in the glass of optical pyrometer

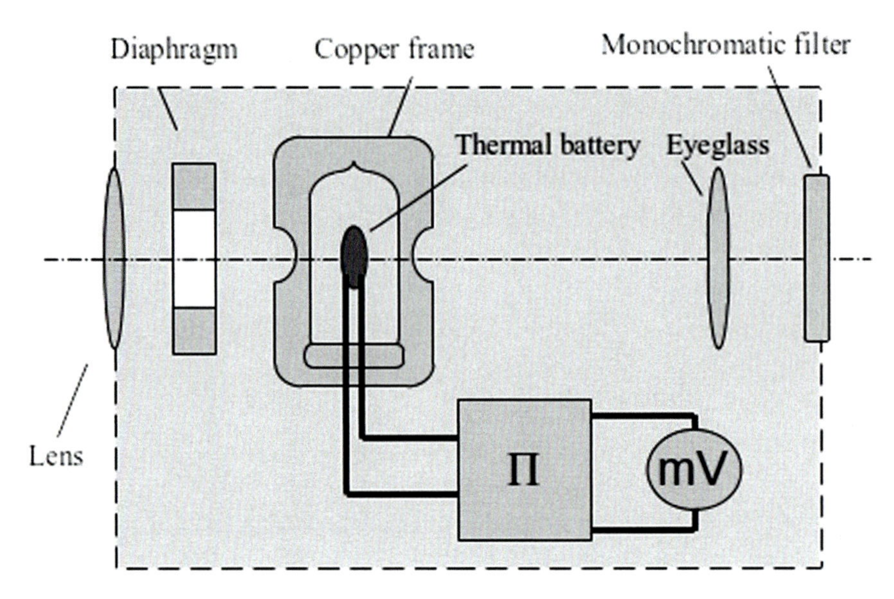

Fig. 2.32 Total radiation pyrometer

in different parts of the spectrum. The point of this approach is that the color of a hot object is related to its temperature. Accordingly, by comparing the intensity of the two spectral components of radiation (for example, red and blue), you can determine body temperature.

Radiation from a heated object is focused by the lens on a shutter—a disk with holes. The first half of the shutter holes are covered with red filters, and the second

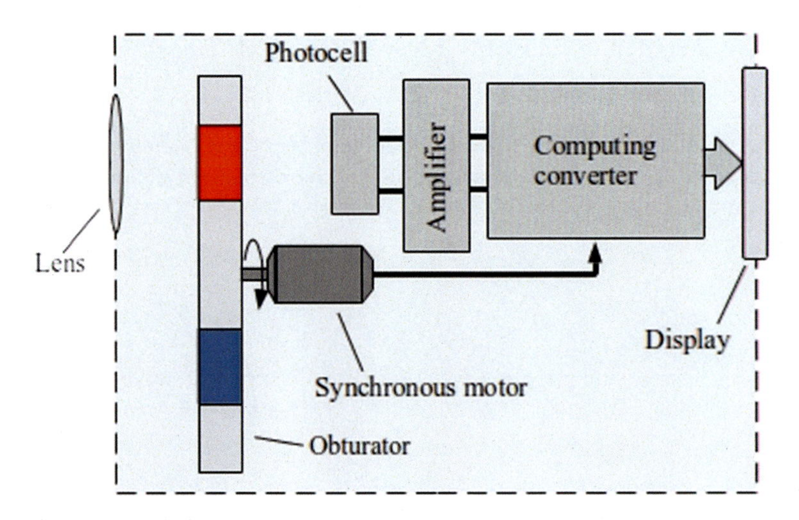

Fig. 2.33 Color pyrometer

half—with blue. The shutter rotates with a synchronous motor. Thus, the photocell is alternately illuminated by red and blue rays, which leads to pulsations of the current generated by the photocell. Ripple parameters are determined by the number of holes in the shutter and its speed.

After amplification, the signal from the photocell is divided according to the ratio of the intensities of the red and blue rays by the computing device, the ratio of their intensity and, accordingly, the temperature is determined.

2.3.6 Measuring Humidity

Humidity is an indicator of the water content in physical bodies or environments. In the welding industry, the humidity of the air in the environment is most often normalized as a parameter that determines the storage conditions of welding and auxiliary materials, the safety of electrical equipment, the conditions of the technological process, working conditions.

Assessment of humidity is carried out, as a rule, by one of two indicators: absolute and relative humidity.

Absolute humidity is defined as the amount of water vapor in 1 m^3 of air (g/m^3).

Relative humidity is defined as the ratio of the current absolute humidity to the maximum possible absolute humidity at a given temperature. Relative humidity is determined as a percentage.

As a rule, the requirements for working conditions or storage conditions indicate the relative humidity and the range of permissible temperatures.

Today, a large number of methods for determining humidity of air, liquid, solid and bulk materials have been developed and used in industry and everyday life. The textbook lists some of the widely used methods.

The **dew point hygrometer** contains a cooled mirror and a thermometer. In the process of measurement, the temperature is recorded at the time of formation of dew on the mirror. The mirror can be cooled by the Loop element, and the moment of dew formation on the surface of the mirror can be captured by a photocell.

Hygrometers also use mechanical materials whose optical or electrical parameters (specific conductivity or dielectric constant) vary depending on their humidity or fogging.

The **psychrometer** contains two thermometers, one of which is dry, the other wet (Fig. 2.34). Humidity is determined from the tables by the difference between the readings of dry and wet bulb thermometers. The lower the humidity, the more intensely the water evaporates and the greater the difference in thermometer readings.

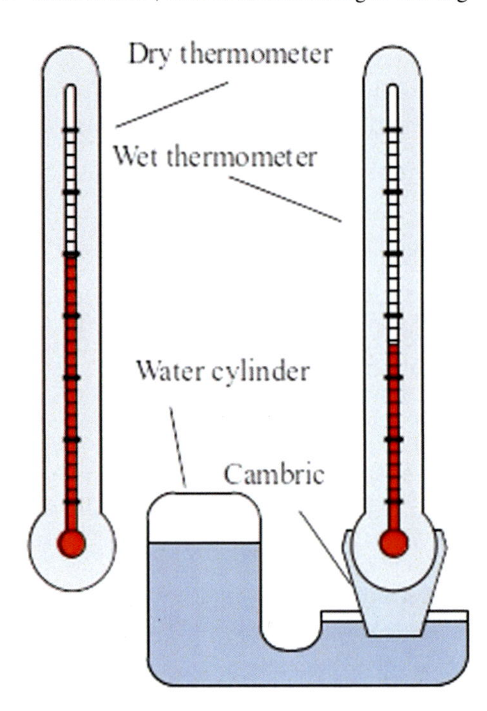

Fig. 2.34 Psychrometer

2.3.7 *Measuring Wind*

Determination of the speed of gas flows (wind, drafts) is necessary to control the conditions of technological processes of arc welding in shielding gases and working conditions.

Devices for measuring wind are called **anemometers**.

Anemometers with impellers and thermoanemometers are the most suitable for measuring wind speed or drafts in the welding zone.

Anemometers with an impeller contain a light wind wheel and a sensor of its rotation speed—optical or based on the Hall effect. The electronic circuit allows us to receive results of measurement in digital format. Such anemometers are sensitive to the position of the impeller relative to the direction of gas flow.

Thermoanemometers use the effect of increasing heat loss by a heated body due to its blowing with colder gas. The thermoanemometer contains an open heating element—a thin metal wire heated by an electric current. Since the resistance of the heating element depends on the temperature, the temperature can be measured fairly accurately by resistance, using a bridge circuit. The low weight of the heating element makes the measuring instrument almost energy-free. The electronic circuit allows us to receive results of measurement in a digital format. It should be noted that the readings of the thermoanemometer are affected not only by the speed of the gas flow, but also its pressure and humidity.

2.4 Calibration, Verification, and Validation of Welding Equipment

Modern quality management systems set requirements for all measuring instruments used to verify product compliance. In addition, a number of requirements for measuring instruments used for commercial accounting, labor safety and the environment are generally set by applicable national legislation. General guidelines for measurement control within the organization are provided by the ISO 10012 [2] standard.

Typically, welding, and related processes can be attributed to technological processes in which the result cannot be easily or economically documented by further control or testing. For such processes, the demonstration of product compliance with the established requirements can be carried out by confirming the compliance of the parameters of the technological process with the regulatory documentation.

Thus, both individual measuring instruments and those built into the process equipment used in the control of the welding process and the control and testing of products must undergo appropriate calibration, verification, or validation procedures [6].

Calibration—an operation by which, under given conditions, the first stage establishes the relationship between the values of the value with the measurement uncertainties that provide standards, and the corresponding readings with the associated measurement uncertainties, and the second stage uses this information to establish ratio to obtain the measurement result from the impressions. That is, it is the process of matching the readings of the measuring instrument (together, if any, with the appropriate software) and the value of the measured physical quantity, considering all the uncertainties that accompany it. Calibration is a procedure for determining the metrological characteristics of measuring instruments.

Verification—providing objective evidence that the object fully meets the established requirements. This can be done on the basis of tests or research, a certain certificate confirming compliance with the requirements. Thus, based on the results of checking the metrological characteristics of measuring instruments located in the field of state metrological supervision, and establishing compliance with the requirements, the authorized state body provides a ***verification certificate***.

However, not always a verified measuring instrument can be used for its intended purpose in a production environment. Thus, a verified metal ruler with a length of 1 m cannot be used to control the size of 0.98 m as indicated in Fig. 2.35.

Validation, attestation—verification, in which the established requirements correspond (are adequate) to the intended use. For example, confirmation that the calibrated voltmeter is suitable for measuring voltage during modulated current welding.

The requirements for the measured values of the technological process or control follow from the technical requirements for the welding procedure specification (WPS). General purpose measuring instruments used in welding sometimes have metrological characteristics that exceed the needs of the technological process.

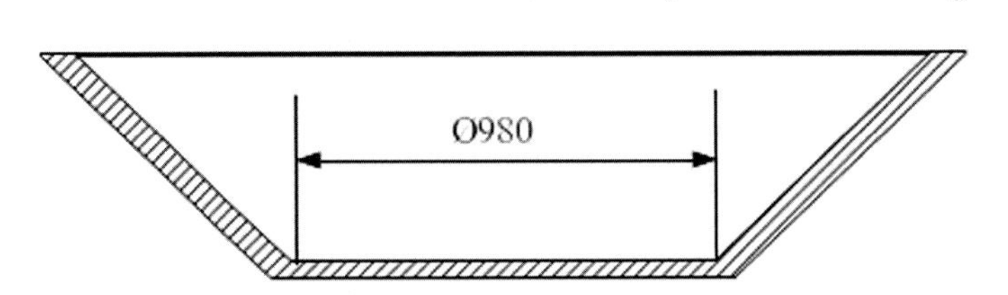

Fig. 2.35 Validation of measuring instruments

Accordingly, standardized calibration, verification or validation procedures may be excessively stringent and not economically justified.

Supervision over the proper condition of measuring instruments and the adequacy of measurement procedures is carried out by the metrological service of the organization.

2.4.1 Periodical Calibration

Calibration, verification, or validation is usually carried out in accordance with metrological supervision legislation, usually once a year. This period can be extended if the reliability and repeatability of the equipment is confirmed. Reduction of the period is necessary in case of deterioration of the equipment, on the recommendation of the equipment developer or the requirements of the customer.

Of course, the calibration procedure of measuring instruments must be carried out at:

- exceeding the calibration period,
- damage or missing seals,
- improper operation of equipment,
- the presence of damage that may affect the work,
- after improper use or external influences that could cause damage (water, temperature, etc.),
- after restoration or repair.

Non-compliant measuring instruments and equipment containing non-compliant measuring instruments must be removed from the production process. After repair, such measuring instruments require confirmation of metrological suitability (calibration).

2.4.2 Confirmation of Metrological Characteristics

Confirmation of metrological characteristics shall be performed for all measuring instruments applicable to welding and related processes according to WPS.

However, in certain circumstances calibration, verification and validation may not be performed.

Thus, calibration, verification and validation may not be performed in the absence of legal or contractual requirements for verification or validation of the relevant processes. This is possible when the result of the process is easy to check with a sufficient level of reliability, for example, in plasma cutting.

In mass or large-scale production calibration, verification and validation of welding equipment and appropriate measuring instruments may be excluded under certain conditions:

- production is controlled by conducting preliminary tests and periodic testing of production samples,
- existing system of statistical quality control of relevant products,
- the production process is stable in the period between tests of samples,
- preliminary tests and sampling of control samples during production are carried out separately for each production line (welding module).

It should be noted that the procedures for calibration, verification and validation of measuring instruments used for appropriate product quality control cannot be abolished.

For serial and unit production, calibration, verification, and validation of welding equipment may be excluded if the following conditions are met:

- Welding procedures are confirmed by tests,
- The same welding equipment is used for the production and manufacture of test specimens,
- A calibrated system of online monitoring of welding parameters is used.

ISO 17662 [6] gives only the minimum requirements for calibration, verification, and validation of measuring instruments in welding and related technologies (Fig. 2.36). Increasing the requirements of the manufacturer, despite the increase in the corresponding costs, provides certain benefits. Among them are increasing the efficiency of process control and, accordingly, their higher stability and productivity.

High accuracy of information about the parameters of the technological process facilitates the transfer of applicable welding technologies to new equipment. The process of manufacturing a welded structure involves the use of measuring instruments at all stages. Common to most welding and soldering processes are workpiece size measurements. The need for calibration of measuring instruments is determined by the requirements for dimensional accuracy of the welded/soldered structure.

Cleaning before welding involves controlling the parameters of the methods by which it is achieved (abrasive cleaning, washing, etching) and the quality of the results. The need for calibration, validation and verification is determined by the quality requirements for cleaning.

Verification of compliance of joints' preparation requires validated measuring instruments. The location of the product in space during welding does not require great accuracy and therefore the calibration of appropriate measuring instruments

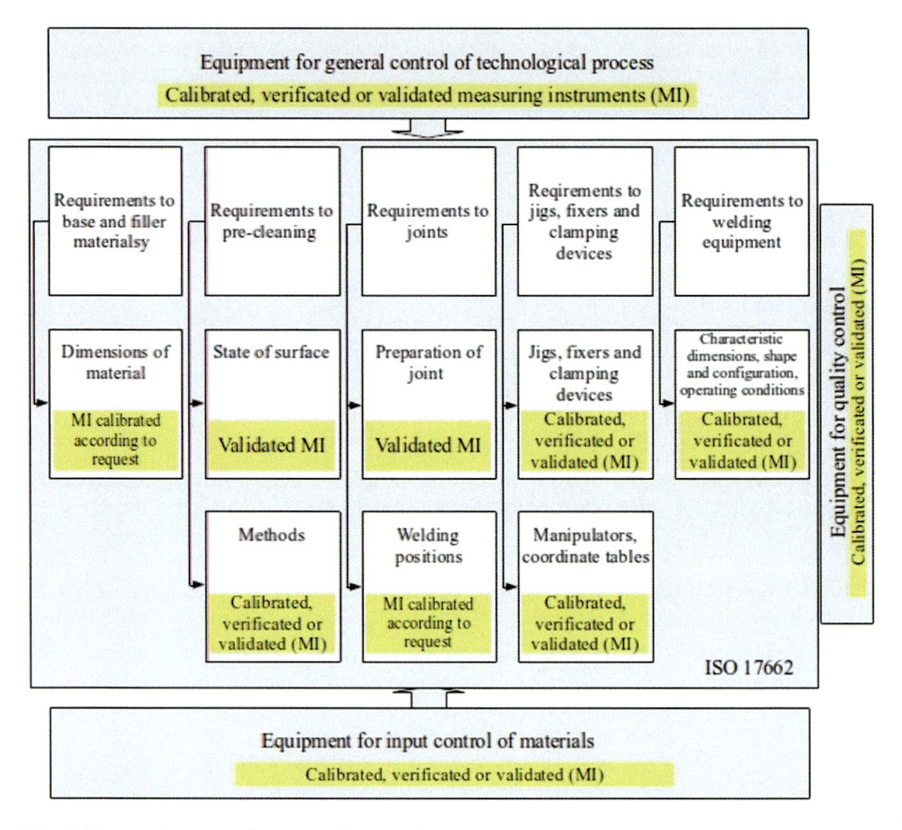

Fig. 2.36 Requirements for calibration, verification and validation

(levels, angles, etc.) require confirmation of their metrological characteristics only in case of damage.

Inspection of auxiliary and main welding equipment requires mandatory confirmation of the metrological characteristics of the relevant measuring instruments.

The need for confirmed metrological characteristics of measuring instruments for a particular process is determined by the requirements for the accuracy of maintaining its parameters specified in the WPS.

Calibration, verification, or validation is not required if the basic parameters of the weld quality can be reliably determined by non-destructive testing methods. However, keep in mind that different non-destructive testing methods only detect defects specific to each method.

Welding is considered a "special process" because the quality of welded joints cannot be fully determined by further inspection and testing of manufactured products and can only be detected during operation of the structure.

Input control of basic and auxiliary welding materials can be carried out both based on certificates and accompanying documentation, and tests. Measuring instruments used in the input control of welding materials are controlled by procedures adopted within the existing quality management system at the organization.

References

1. ISO/IEC Guide 99:2007 International vocabulary of metrology—Basic and general concepts and associated terms (VIM)
2. ISO 10012:2003 Measurement management systems—Requirements for measurement processes and measuring equipment
3. ISO 13916:2017 Welding—Measurement of preheating temperature, interpass temperature and preheat maintenance temperature
4. IEC 60584-1:2013 Thermocouples—Part 1: EMF specifications and tolerances
5. IEC 60584-3:2021 Thermocouples—Part 3: Extension and compensating cables—Tolerances and identification system
6. ISO 17662:2016 Welding—Calibration, verification and validation of equipment used for welding, including ancillary activities

Chapter 3
Imperfections and Acceptance Criteria

3.1 Types of Weld Imperfections. ISO 6520-1

3.1.1 Identification of Imperfections

Imperfection—a cavity in the welded joint or deviation from required geometry.

According to ISO 6520-1 [1] welding imperfections are classified into 6 basic groups (in imperfection identification a basic group is depicted as the first number in numerical ID of imperfection):

(1) cracks,
(2) cavities,
(3) solid inclusions,
(4) lack of fusion and penetration,
(5) imperfect shape,
(6) miscellaneous imperfections.

Within each group, imperfections are divided into subgroups and types:

- subgroup—by orientation or mass character of imperfections (in imperfection, subgroup ID is depicted as the second and third numbers),
- types—by location in zones of welded joint or by mass character (in imperfection, type ID is depicted as the fourth number).

Numerical identification of imperfection is written in the following way:

Imperfection ISO 6520-1- _ _ _ _ *—where '_' are replaced with numbers according to the following:*

- *group—first number,*
- *subgroup—second and third numbers,*
- *type—fourth number.*

For example, numeric identification "**Imperfection ISO 6520-1-1023**" means (see Fig. 3.1).

S. Fomichov et al., *Quality Management in Welded Fabrication*,
https://doi.org/10.1007/978-3-031-34800-6_3

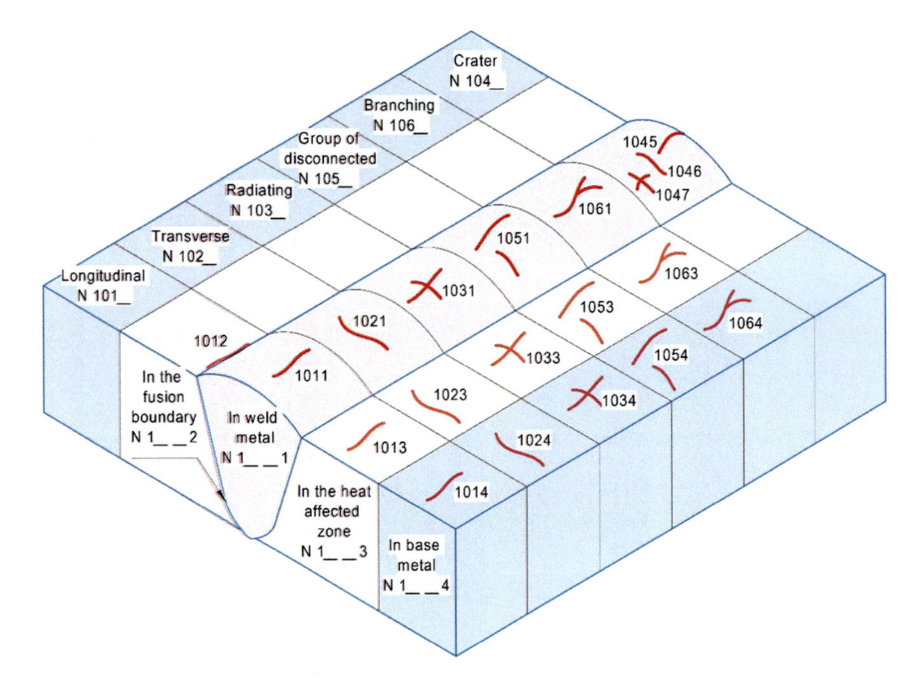

Fig. 3.1 Types of cracks by orientation relative to weld axis and location in welded joint

- Group: **1:** crack,
- Subgroup: **02:** transverse,
- Type: **3:** in the heat-affected zone.

3.1.2 Cracks

(1st group of imperfections)

Crack—a disruption of inter-atomic bonds of metals' crystal cell with forming of free surfaces (crack shores) as a result of:

- constant, variable and impact loads (strength factor),
- changes of metal structure (metallurgical factor).

Uneven heating and cooling of weld metal and metal in the HAZ during welding lead to variation of both strength and metallurgical factors.

Usually crack starts forming from stress concentrator.

Cracks are considered to be the most dangerous imperfections in welded joints.

ISO 6520-1 [1] classifies cracks into two categories:

(1) orientation and location—basic classification (see Fig. 3.1),
(2) origin causes—additional classification.

By orientation cracks are divided into the following subgroups:

101_—longitudinal—cracks parallel to the weld axis,

102 _—transversal—cracks perpendicular to the weld axis,

103 _—radiating—cracks which radiate from one point,

105 _—disconnected cracks—group of cracks which are not connected to each other and are oriented in different directions,

106 _—branching—group of connected cracks emanating from one common crack.

By location cracks are divided into four categories:

_ _ _ 1—in the weld metal,

_ _ _ 2—in the fusion zone,

_ _ _ 3—in the HAZ,

_ _ _ 4—in the base metal.

Cracks located in the weld crater are classified as a separate subgroup 04:

1045—longitudinal crater cracks,

1046—transversal crater cracks,

1047—radiating crater cracks.

In addition, basic classification includes microcracks 1001—cracks visible under the microscope only (not shown on Fig. 3.1).

In quality assurance and quality control in welding fabrication the classification of cracks by their origin is also very important. This classification with letter designation is given in ISO 6520-1 [1] (Appendix):

Ea hot crack,

Eb solidification crack,

Ec liquation crack,

Ed precipitation induced crack,

Ee age hardening crack,

Ef cold crack

Eg ductility-dip crack,

Eh shrinkage crack,

Ei hydrogen-induced crack,

Ej lamellar tearing,

Ek toe crack,

El ageing induced crack (nitrogen diffusion crack).

When identifying cracks, letter designation is added to the numerical one (see Sect. 3.1.1).

Hot cracks—brittle intergranular disruption of weld metal or metal in the HAZ which originates either in solid-liquid state during crystallization or in solid state within high temperatures during the stage of predominant development of intergranular deformation.

Hot cracks originate as a result of a combination of two physical effects:

(1) Brittleness temperature ranges (BTR)—ranges of temperature during weld metal crystallization and cooling of welded joint in which plastic characteristics of metal spasmodically reduce to the level of minimal plasticity. Possibility

of crack originating depends on BTR value and cooling temperatures (rates) within which BTR appears. Different steels and alloys may have from one to three BTRs.

(2) Presence of tensile strains during cooling of welded joint. They appear as a result of uneven thermal plastic deformation of different zones of welded joint. Crack originates when the rate of high-temperature welding deformation is greater than the ability of metal to deform in a particular brittleness temperature range.

Depending on temperature of the brittleness temperature range and factors leading to reduction of plasticity in BTR, hot cracks are divided into three categories (see Fig. 3.2).

(1) solidification cracks—originate in BTR_1,
(2) liquation cracks—originate in BTR_2,
(3) precipitation induced cracks—originate in BTR_3.

Hot cracks originate under high temperatures and are intergranular, therefore their fissures are dark in color (due to oxides on the surface) and their tips are round.

Solidification crack (Fig. 3.2a)—are hot cracks which originate in solid–liquid state of metal under temperatures higher than solidus temperature (first brittleness temperature range BTR_1).

Under temperatures close to that of liquidus, the deformation ability of metal is high due to great amount of liquid phase. Metal deforms with relative movement of solid zones and circulation of liquid among them.

With further cooling of the weld the amount of liquid phase reduces. Crystallites appear in contact with each other, and liquid circulation becomes limited. Deformation ability of the weld metal reduces to its minimum. Temperature related to this state is BTR_1 border high.

When metal is deformed in solid–liquid state closer to the solidus temperature the deformation affects zones of crystallite contacts. If the total deformation ability of these zones in any section is lower than the rate of weld metal deformation, solidification crack originates (usually, these sections are oriented perpendicular to the weld metal deformation).

With further cooling to the lower border of BTR_1 temperature the deformation affects the whole volume of solidified metal. As a result, the deformation ability of the metal increases up to its maximum and crack originating conditions disappear.

As far as the mechanism of plasticity loss in BTR_1 is present during all crystallization processes, solidification cracks can appear in all steels and alloys. Probability of solidification crack origination depends on relation between rate of deformation and deformation ability of the weld metal. Therefore, activities aimed at solidification cracks (as well as other hot cracks) prevention can be performed in two ways:

(1) Reduction of deformation rate during crystallization:

- pre-heating of base metal before welding and additional heating of weld during welding (up to 250–450 °C),

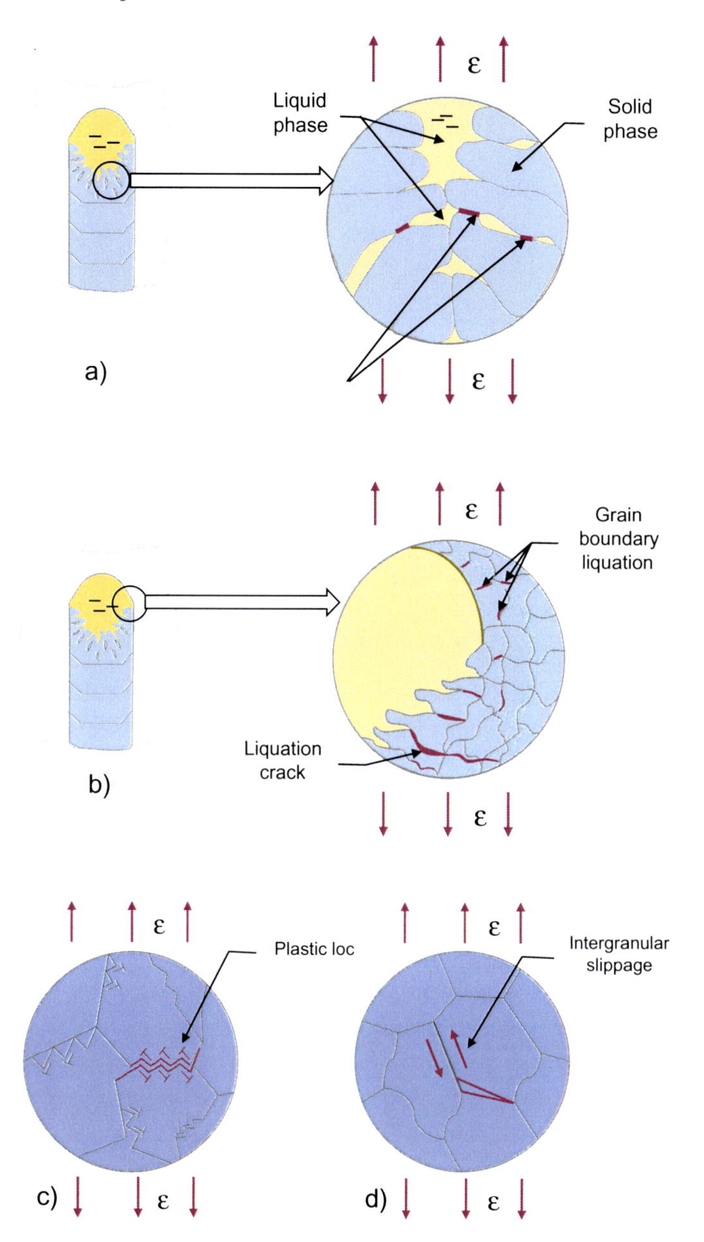

Fig. 3.2 Mechanisms of hot cracks origination: **a** solidification crack (BTR_1), **b** liquation crack ($BTRX_2$), **c** localization of plastic deformations on border zones (BTR_3), **d** relative slipping of grains and crack origination (BTR_3)

- reduction of structure stiffness in the welding zone (by assigning fit-up and welding sequence or by welding with discontinuous welds),
- reduction of welding speed and welding current with increase of voltage and gap.

(2) Increase of deformation ability of weld metal:

- reduction of content of impurities (sulfur, phosphorus, etc.) in base metal and in weld metal.

Liquation crack (see Fig. 3.2b)—are hot cracks originating usually in the fusion zone and growing both into the weld metal and into the HAZ. Liquation cracks (similarly to solidification ones) originate due to low deformation ability in the zone of welded joint in relation to rate of deformation in this zone. However, there are some differences from solidification cracks, such as:

- deformation ability decreases due to segregation of impurities on the grain borders because of diffusion processes during heating and subsequent border melting under welding heating,
- process is going on under temperatures lower than solidus (solid metal state) after primary crystallization is finished (second brittleness temperature range BTR_2).

Precipitation induced crack—are hot cracks originating in solid metal completely without participation of liquid phase in the third brittleness temperature range (BTR_3). By their nature they are ductility-dip cracks.

Precipitation induced cracks usually appear in high-alloyed heat-resistant austenitic steels and nickel alloys.

Ductility-dip crack—originates in BTR_3 because of intensive diffusion of atoms during secondary crystallization and grain border migration.

When fine-grained intermetallic and carbonitride phases diffuse from solid, the solution grains become embrittled and hardened. Dispersion hardening of grains leads to localization of plastic deformations in border zones (see Fig. 3.2c), relative slipping of grains and crack origination (see Fig. 3.2d).

Ductility-dip cracks usually appear in austenitic steels and nickel alloys.

Age hardening crack—a variation of reheat cracks induced by diffusion of excessive phases (usually, carbon) from supersaturated solid solution of metals and alloys during heat-treatment and plastic deformation. As a result, hardness and strength increase with decrease of ductility and plasticity.

During heat-treatment on its cooling stage, different parts of welded structure cannot uniformly change their size under effect of temperature due to:

- different part stiffness (e.g., outside shells of cylindrical pressure vessel have higher stiffness in comparison to inner ones because they are connected to bottoms),
- differences in heat conductivity of structural elements (flanges, fittings, support units cool down faster than bigger elements),

- differences in heat sink in central and peripheral zones of the oven.

All the above leads to formation of temperature deformations and tensions. Together with plasticity loss they make crack origination possible.

Shrinkage crack—are hot cracks originating during crystallization usually of small metal volumes in a hard outline because of shrinkage forces. Longitudinal, transverse, and radiating crater cracks as well as cracks appearing during resistance spot welding are shrinkage ones.

Cold crack—is a brittle disruption of the HAZ or less often—of the weld metal. It happens during cooling usually to temperatures lower than 200 °C. Cold crack can originate up to a couple of days after welding was performed. Cold cracks appear if two conditions are met:

(1) Characteristics of plasticity of the base metal are reduced due to:

   - effect of alloying elements,
   - effect of diffusion hydrogen,
   - low temperatures (lower than −300 °C), which also reduces plasticity of the base metal and contributes to brittle crack growth.

(2) Energy condition—energy of elastic deformation freed during crack growth should be higher than increment of full surface energy of crack borders. By energy condition the crack is being "fed" by:

   - residual welding tensions,
   - stress concentrators, imperfections at the first place—simplify crack origination (that's why cold cracks formation is prevented on the weld toe angle and in the weld root),
   - fatigue or impact load.

The possibility of cold cracks in steels is evaluated by carbon equivalent C_{eq}. It is calculated as a total alloying elements' percentage in steel with corresponding coefficients:

$$\begin{aligned} C_{eq} &= C + Mn/6 + Cr/5 + Mo/4 + V/14 \\ &\quad + Ni/15 + Cu/15 + Nb/4 + Ti/4 + P/2 + B * 5 \end{aligned} \tag{3.1}$$

If $C_{eq} > 0.45°$—cold cracks will form.
If $C_{eq} < 0.25°$—cold cracks will not form.
Due to origination under low temperatures cold cracks fracture is shiny and without traces of high-temperature oxidation.
Cold cracks can be prevented by:

(1) Control of mechanical factor—reduction of welding tensions:

   (a) Pre-heat of base metal before welding and additional heating of weld metal during welding (up to 250–450 °C).

(b) Heat treatment of welded joints:

- High tempering—heating to 650–750 °C, holding for 1–5 h and slow cooling in the oven. Hardness is reduced and unified, plasticity and toughness increase. Residual welding stresses are reduced by 70–80%.
- Low tempering—heating to 3000 °C, holding for 1–5 h and slow cooling in the oven. Rate of structural stresses is reduced, diffusion-mobile hydrogen is removed.
- Normalizing—heating to 900–950 °C, holding during a couple of minutes and air cooling. Fine-grained metal structure is formed in the weld, hardness, plasticity and toughness increase, welding tensions are reduced.

(2) Control of metallurgical factor—application of ductile filler materials, for example, austenitic welding wires.

Hydrogen-induced crack originates and grows as a result of:

- Metal embrittlement due to diffused hydrogen preventing migration of dislocations and, as a result, preventing plastic deformation.
- Molization of diffused hydrogen (formation of hydrogen molecules from two hydrogen ions) in inner cavities. The hydrogen molecule is bigger in size and cannot move inside the metal. Microcavity becomes a "trap" for molecular hydrogen. As a result, high inner pressure appears, and cavity starts to develop.

Main activities to prevent hydrogen-induced cracks are aimed at removing hydrogen from the weld zone:

- drying of fluxes and covered electrodes prior to welding,
- cleaning of welding wire from rust and oils,
- cleaning of edges prior to welding,
- protection of welding zone.

Lamellar tearing—are cracks developing in the base metal of welded structures made of plates and pipes as a result of combined effect of metallurgical and mechanical factors (see Fig. 3.3).

Metallurgical factor defines crack origination and includes:

- non-metallic inclusions (primarily, sulfides) in slabs and pipe shells, which during rolling are transformed into tearing,
- rolling textures—elongated grains (and grain borders with impurities) oriented along the rolling direction.

Mechanical factor defines crack development. It includes effect of tensions transversal to rolling direction:

- temporary welding tensions caused by welding heating cycle,
- residual welding tensions,

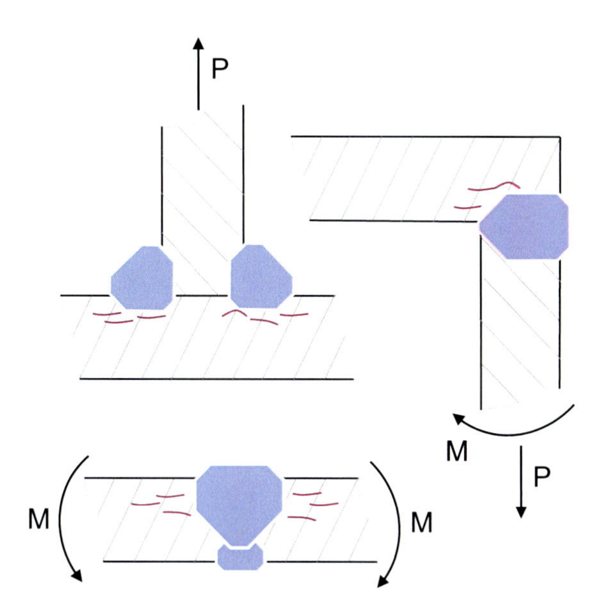

Fig. 3.3 Lamellar tearing in welded joint

- service loads.

Development of lamellar tearing is stepped and is going on due to growth and merging of tearing from different layers.

Lamellar tearing usually appears in angle joints and T-joints in structures made of high-strength steels.

Toe cracks emanate from the border of weld reinforcement as a stress concentrator. They develop into the base metal under effect of residual welding stresses and external loads. Such cracks appear in multilayer welds, including those joining flanges to vessels as well as in welds with galvanic coatings. In this case cracks are initiated by disruption of coating along the weld border (Fig. 3.4).

Ageing induced crack (nitrogen diffusion crack)—is a slow disruption of metal caused by changes in its mechanical characteristics as a result of diffusion, primarily of nitrogen.

Ageing causes metal embrittlement due to formation of excessive nitride phases ($Fe_{16}N_2$ *or* Fe_4N) from ferrite. This process contributes to crack development.

Ageing is typical for low-carbon steels (less than 0.25%) and is usually caused by:

- high-rate cooling under 650–700 °C—thermal ageing,
- plastic deformation happening under temperatures lower than that of recrystallization (ageing can be spotted within 15–16 days)—deformation ageing.

Solving technological tasks is based on principal understanding of crack nature. Generally, they can be classified into three basic groups (Fig. 3.5):

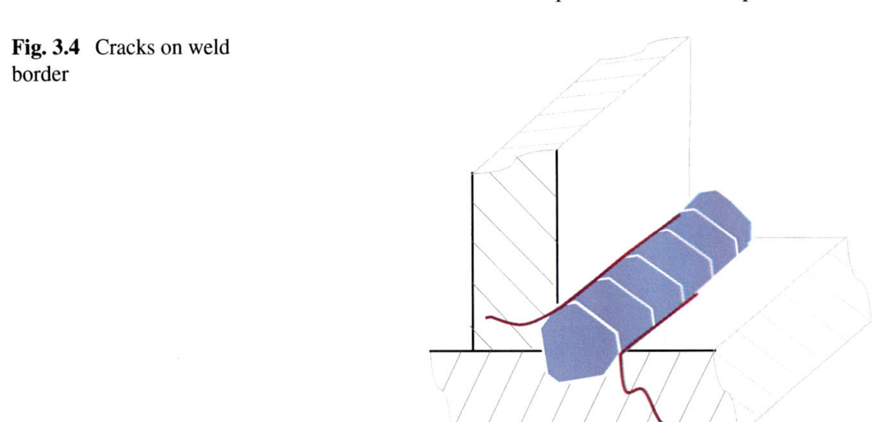

Fig. 3.4 Cracks on weld border

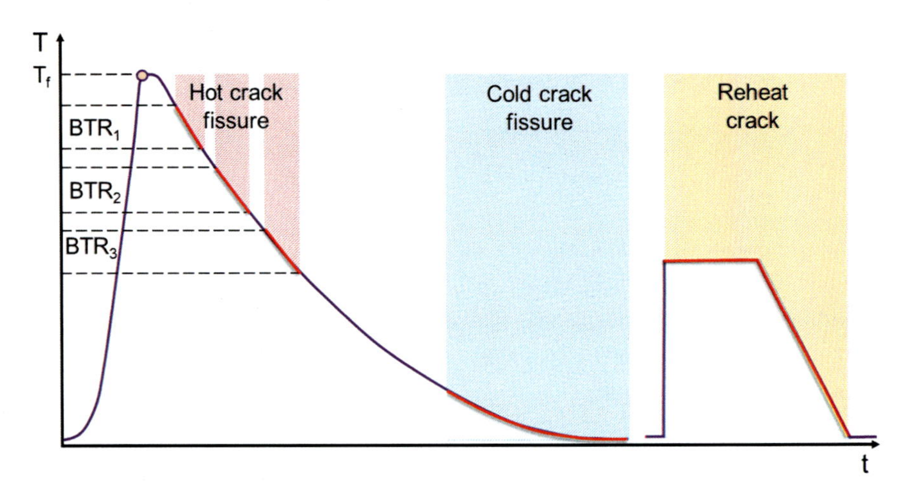

Fig. 3.5 General groups of cracks in welding

(1) **Hot cracks**—originating either in solid–liquid state during crystallization (BTR_1), or in solid state under high temperatures of stage of development of intergranular deformation (BTR_2 and BTR_3). Hot cracks are caused by a combination of two physical effects:

- Step loss of plasticity of metal in brittleness temperature ranges,
- Temporary welding tensile deformations present during weld cooling. Crack originates if rate of high-temperature welding deformation exceeds deformation ability of metal in the particular brittleness temperature range.

(2) **Cold cracks**—originate in temperatures under 200 °C. Cold cracks are caused by combination of two conditions:

- Reduction of plastic characteristics of the base metal caused by effect of alloying elements and (or) diffused hydrogen effect.
- Energy condition—increment of energy for crack development caused by residual stresses. In this case stress concentrators, imperfections in particular, simplify crack origination.

(3) **Reheat cracks**—originate because of heat treatment and plastic deformation after welding. Reheat cracks are caused by excessive phases (often by carbides) in solid metals and alloys. As a result, hardness increases, but ductility and plasticity decrease.

General crack preventing actions in welding include:

(1) Preventing embrittlement and (or) increase of plasticity of weld metal and HAZ.
(2) Reduction of deformations and tensions during welding, reduction of residual welding tensions.
(3) Minimizing stress concentrators in welding zone, primarily welding imperfections.

3.1.3 Cavities

(2nd group of imperfections)

Cavities are imperfections appearing in the weld. There are two subgroups of cavities—gas cavities (20**1**_) and shrinkage cavities (20**2**_)—Fig. 3.6.

201_—**Gas cavity**—are cavities formed by captured gas freed during crystallization.

Pores are main type of gas cavities. They have spherical shape which can be: gas pore (201**1**), uniformly distributed porosity (201**2**), clustered (localized) porosity (201**3**), linear porosity (201**4**) and surface pore (201**7**).

Two other types of gas cavities are:

- elongated cavity (201**5**),
- worm-hole (201**6**).

Mechanism of formation of gas cavities, pores at the first place, includes the following stages:

(1) Gases get into the welding zone and dissolve in the liquid metal of the welding pool. Such gases are:

   (a) hydrogen—from water contained in fluxes, electrode coatings, rust, and oxides on edges to be welded and on welding wire, in shielding gases and in the air,

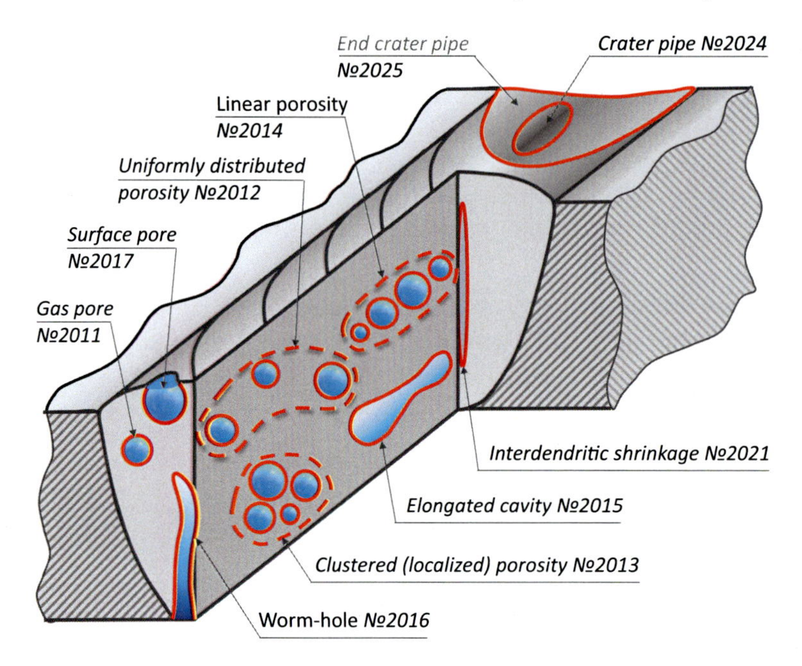

Fig. 3.6 Cavities in the weld

(b) nitrogen—from the air,
(c) carbon dioxide—forms during burn-out of dirt (oils, etc.) on edges, on welding wire and electrodes.

(2) When temperature of the welding pool reduces, solubility of gases in metal reduces as well. As a result, gas starts to free up as bubbles.
(3) Bubbles partially flow to the surface where they are removed from the welding pool. The rest of bubbles remains in the weld after crystallization is finished as pores.

Activities preventing gas cavities are aimed at preventing gases from getting into the welding pool and at easing their evolution and removal during crystallization:

- good protection of welding pool, including the root,
- protection of welding materials from humidity (drying of coated electrodes, flux, refining of shielding gases),
- cleaning rust, oxides, and oils from edges to be welded and from welding wire before welding,
- increasing of time during which metal is in liquid state, for example, by reducing welding speed,
- reduction of contact surface between arc and air to prevent getting nitrogen into the welding zone, for example by reduction of arc length,
- improvement of welder's qualification,

- reduction of content of surface-active alloying elements, such as silicon, titanium as well as sulfur in the base metal (surface-active elements slow down hydrogen desorption and worsen degassing of the welding pool),
- application of external magnetic fields to stir the metal in the welding pool.

202 _—**Shrinkage cavity**—are the cavities which are formed due to shrinkage during crystallization.

Shrinkage cavities, as a rule, are located on places of arc disruption and are caused by lack of metal during crystallization.

Activities to prevent shrinkage cavities are:

- application of lead bars,
- improvement of competency of MMA, MIG/MAG and TIG welders,
- correction of welding parameters for SAW.

3.1.4 Solid Inclusions

(3rd group of imperfections)

Solid inclusions are imperfections of the weld. They are classified by composition. There are 4 subgroups of solid inclusions (see Fig. 3.7):

301 _—slag inclusions (SAW, MMA, FCAW, ESW welding),

302 _—flux inclusions (SAW and MMA welding),

303 _—oxide inclusions (when welding aluminum and its alloys),

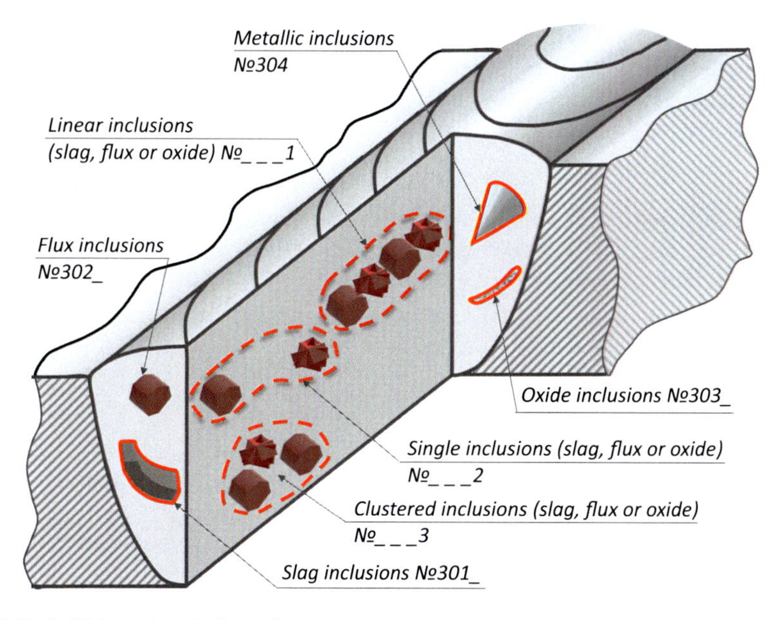

Fig. 3.7 Solid inclusions in the weld

304 _—metallic inclusions (TIG welding at the first place).

In addition, by mass factor and location slag, flux and oxide inclusions are divided into:

3 _ _ **1**—linear,
3 _ _ **2**—single,
3 _ _ **3**—clustered.

Slag inclusions are pieces of slag stuck in the weld:

- when the slag was not completely removed after previous welding run (mainly from the weld reinforcement) in case of multi-run welding,
- when molten slag was not brought to the surface of the welding pool until crystallization finished (this happens when molten slag has high density, toughness, or high melting temperature).

Activities preventing slag inclusions are aimed at preventing pieces of slag in the welding pool and removing them from the welding pool:

- using fluxes and electrode coatings which provide good removability of slag,
- reduction of welding speed,
- increase of welding current.

Flux inclusions—are flux granules or pieces of electrode coating which got into the welding pool, but have not molten there, or were not moved to the surface until crystallization was finished. This may be caused by:

- flux being located outside of the arc affected zone, e.g., in lower part of the groove (in the root zone),
- electrode coating was melting unevenly, starting to crush and its pieces appeared in the welding pool,
- flux was not fully removed from the previous run-in case of multi-run welding,
- high welding speed,
- low welding current.

Oxide inclusions appear primarily when welding aluminum and its alloys. If edges or surface of the previous run were not cleaned properly, aluminum oxide film gets into the welding pool and oxide inclusion is formed due to oxide's high melting temperature.

In addition, oxide inclusions can be formed when welding steels if solubility of impurities is reduced due to decrease of temperature during crystallization.

Oxide inclusions may be prevented by:

- proper cleaning of edges to be welded and of the previous run surface in case of multi-run welding,
- increase of time during which weld metal is in liquid state, for example, by reduction of welding speed.

Metallic inclusions depending on metal composition are divided into tungsten, copper and others.

3041 *tungsten inclusions*—the most common metallic inclusions. They are pieces of non-consumable tungsten electrode for TIG welding. Tungsten has a high melting temperature and is not melting during welding. However, it is very slowly evaporating under high temperature. If evaporation and, as a result, electrode consumption is uneven, for example when electrode's tip is too sharp, the tip may disrupt and fall into the welding pool.

To prevent tungsten inclusions:

- it is necessary to use proper shape electrodes,
- arc should be stroked on low voltage to prevent electrode overheating,
- welding should be performed on reverse polarity,
- welding current should be reduced.

3042 *copper inclusions*—are formed as a result of copper segregation when it appears in the welding pool, usually when using copper-covered welding wire. It is /////important to mention that main part of copper dissolves in iron. During crystallization copper forms solid compound with iron which does not lead to the imperfection.

3043 *other metallic inclusions*—are formed as a result of random metals appearing in the welding pool and differences between their melting temperature with that of the weld metal.

3.1.5 Lack of Fusion and Penetration

(4th group of imperfections)

Lack of fusion (401_)—joint missing due to lack of fusion (see Fig. 3.8):

401**1**—between weld metal and base metal edges,

401**2**—between beads in case of multi-run welding,

401**3**—between weld metal and base metal edges in the weld root zone.

Lack of fusion is caused by:

- improper groove shape,
- deviations of the welding arc and, as a result, of the welding pool from edges,
- impurities on the edges, including oxide films,
- improper previous runs surface preparation in case of multi-run welding,
- low welding current,
- high welding speed,
- low qualification of welder.

Lack of penetration (40**21**)—joint missing as a result of inability of molten metal to get to the weld root (see Fig. 3.8).

Lack of penetration is caused by:

- low welding current (main cause),
- high welding speed,

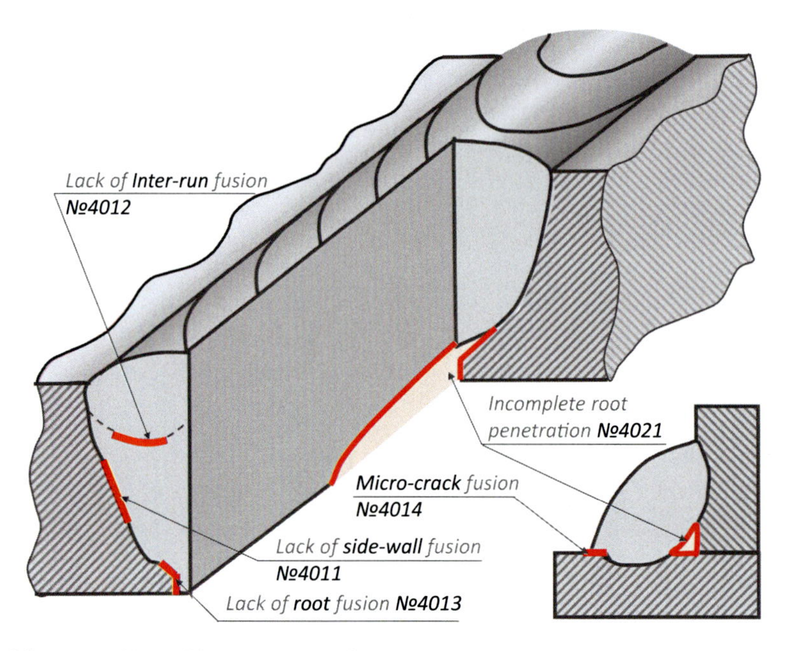

Fig. 3.8 Types of lack of fusion and lack of penetration in the welded joint

- improper groove shape.

3.1.6 Imperfect Shape

(5th group of imperfections)

Imperfect shape group includes 19 subgroups of imperfections:

(1) **501_—undercut** (Fig. 3.9). Depending on characteristics and location, undercuts are classified into five types:

- 501**1**—continuous undercut,
- 501**2**—intermittent undercut,
- 501**3**—shrinkage grooves,
- 501**4**—inter-run undercut,
- 501**5**—local intermittent undercut.

Undercuts are caused by:

- poor heat sink from edges in case of different thickness of welding parts (e.g., T-joints) or in case of welding thin plates,
- high voltage,
- high welding speed,

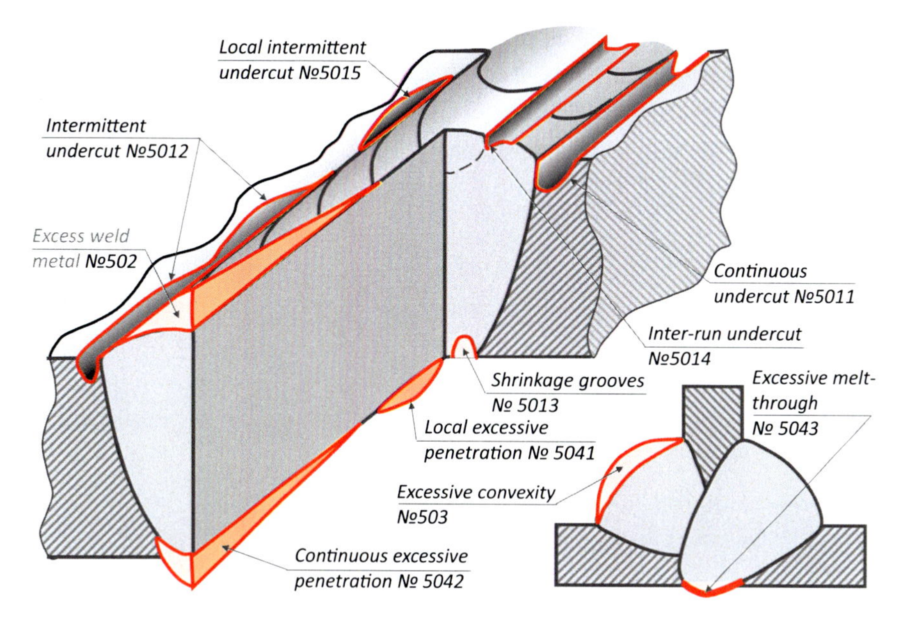

Fig. 3.9 Undercuts, excessive convexity, and excessive penetration in the weld

- shielding gas content in case of MAG welding (good results are obtained when applying mixes containing inert gas),
- improper electrode angle,
- big transversal electrode oscillations in case of MMA welding.

(2) **502—excess weld metal** (Fig. 3.9)—excess of deposited metal on the face of the butt weld.

Excessive convexity is caused by:

- low welding speed (main cause),
- high electrode wire feed speed,
- low qualification of the welder,
- improper shielding gas (using gas mixtures reduces convexity, for example, 80% Ar + 20% CO2 mix for MAG),
- back-angle welding structures with thin elements,
- improper groove shape (narrow groove) in case of single-run welding.

(3) **503—excessive convexity** (Fig. 3.9)—excess of deposited metal on the face of the fillet weld.

Excessive convexity in the fillet welds is caused by the same factors as that in the butt weld (see **502**).

(4) **504_—excessive penetration** (see Fig. 3.9)—excessive deposited metal in the weld root. Depending on continuity and characteristics, these imperfections are classified into three types:

- 504**1**—local excessive penetration,
- 504**2**—continuous excessive penetration,
- 504**3**—excessive melt-through.

Excessive penetration is caused by:

- high welding current (main cause),
- big gap between plates to be welded,
- uneven gap along the weld.

(5) **505_—incorrect weld toe** (see Fig. 3.10)—can appear as wrong angle or wrong radius:

- 505**1**—incorrect weld toe angle,
- 505**2**—incorrect weld toe radius.

Incorrect weld toe is caused by:

- low welding speed (main cause)
- low qualification of the welder,
- improper shielding gas (MAG welding with gas mixtures reduces weld reinforcement and changes the weld toe),
- back-angle welding structures with thin elements.

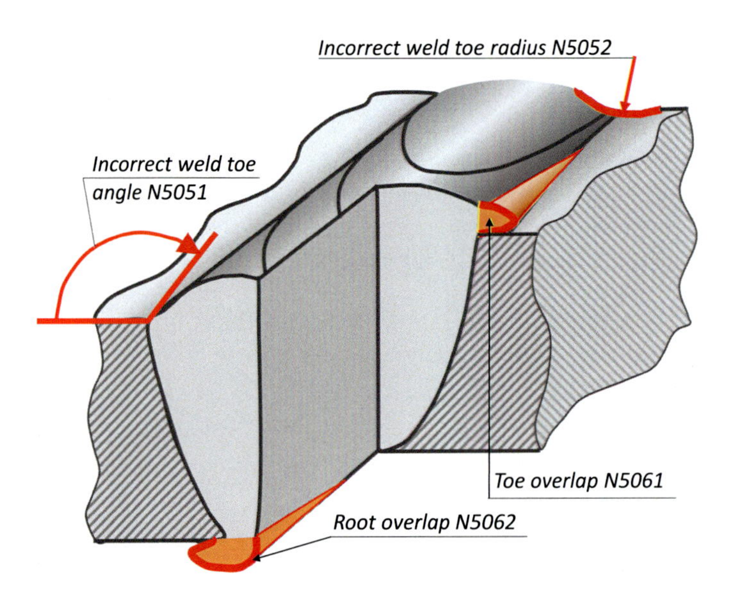

Fig. 3.10 Incorrect weld toe and overlap

(6) **506_—overlap** (see Fig. 3.10)—excess of deposited metal on the base metal surface without fusion between them. Depending on location, overlaps are classified into two types:

- 506**1**—toe overlap,
- 506**2**—root overlap.

It should be mentioned that excessive convexity (5**02**) appears as excess of deposited metal as well. The difference between these imperfections is the excess metal location:

- excessive convexity—on the weld and with fusion,
- overlap—on the base metal without fusion.

Overlaps frequently appear in fillet welds and butt welds welded in horizontal position due to leakage of welding pool molten metal on the edges of the base metal.

Overlaps are caused by:

- improper welding position,
- low qualification of the welder,
- excess of filler metal because of low welding speed or high wire feed speed,
- lack of heating of the base metal because of, for example, low heat input,
- transversal arc deviations, wrong electrode angle,
- slag on the edges to be welded which prevents fusion,
- low ductility of the molten flux,
- improper slider pressing in case of electro-slag welding.

(7) **507_**—linear misalignment (see Fig. 3.11)—misalignment between two elements to be welded that have parallel surfaces located in different planes. Depending on welded structure geometry, linear misalignments are classified into two types:

- 507**1**—linear misalignment between plates,
- 507**2**—linear misalignment between tubes.

Linear misalignment is mainly caused by ineffective jigs and fixtures, that do not provide good fixation of plates and tubes.

(8) **508—angular misalignment** (see Fig. 3.11)—misalignment between two elements to be welded with non-parallel surfaces or surfaces located under wrong angle.

Angular misalignment is caused by:

- ineffective jigs and fixtures which do not provide proper fixation of plates and tubes (main cause),
- transversal shrinkage during root welding and welding of first runs in case of multi-run welding.

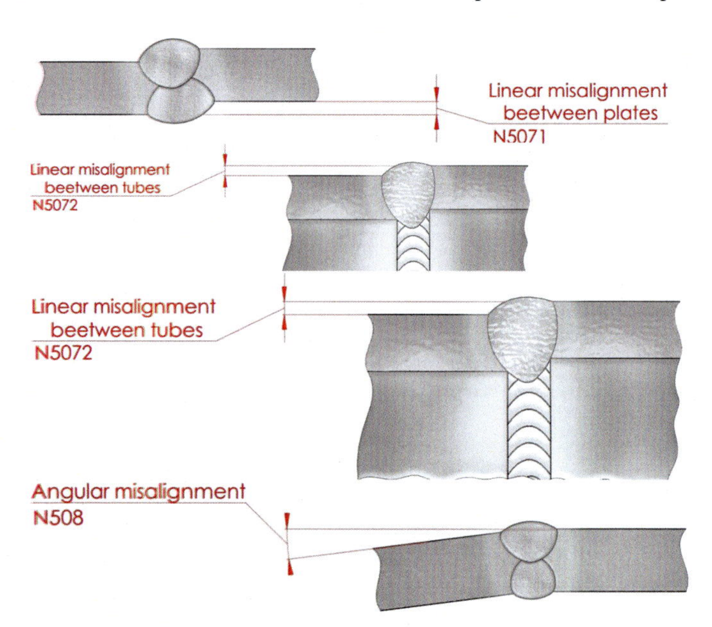

Fig. 3.11 Linear and angular misalignment

(9) **509_—sagging** (see Fig. 3.12)—leakage of the metal due to gravity force. Depending on position and weld type, sagging is classified into four types:

- 509**1**—sagging in the horizontal position,
- 509**2**—sagging in the flat or overhead position,
- 509**3**—sagging in a fillet weld,
- 509**4**—sagging at the edge of the weld.

Unlike overlap (506_) sagging is not caused by excess of deposited metal.
Sagging is caused by:

- disadvantageous welding position (sagging in horizontal position),
- big gap (sagging in flat position),
- electrode deviation to the vertical part (fillet weld sagging)—this results in significant heating, melting and leakage of metal on the horizontal plate,
- low qualification of the welder,
- low ductility of molten flux.

(10) **510—burn-through** (see Fig. 3.13)—leakage of the welding pool with formation of a thorough hole in the weld.
Burn-through is caused by:

- high welding current for the given metal thickness (main cause),
- low welding speed,
- low qualification of the welder,

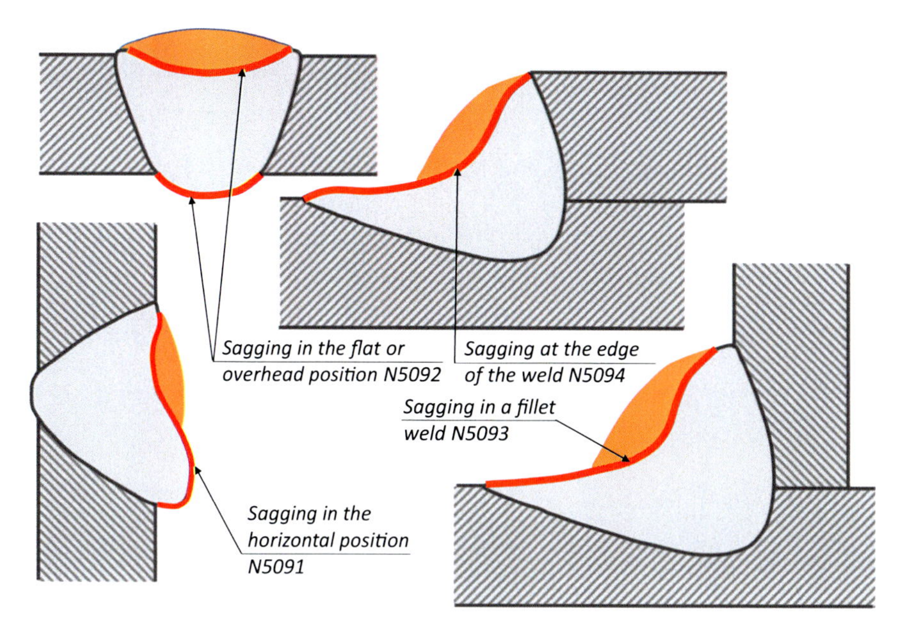

Fig. 3.12 Sagging in the weld

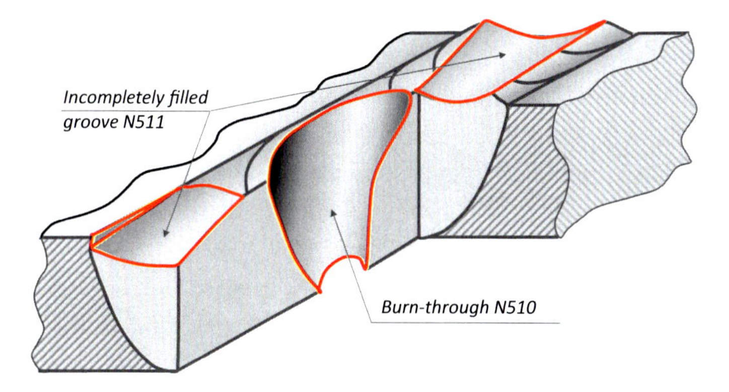

Fig. 3.13 Burn-through and incompletely filled groove

- improper pressing of copper or flux backing,
- big gap between parts to be welded,
- small root face,
- improper shielding gas or flux.

(11) **511—incompletely filled groove** (see Fig. 3.13)—continuous or intermittent, depending on the weld surface due to lack of deposited metal.

Incompletely filled groove is caused by:

- improper groove shape (main cause),
- low qualification of the welder.

(12) **512—excessive asymmetry of the fillet weld** (see Fig. 3.14)—unequal leg length.

Excessive asymmetry of the fillet weld is caused by:

- improper positioning of parts to be welded,
- asymmetry of the heat sink caused by fillet weld geometry, as a rule by different thicknesses of parts.

(13) **513—irregular width** (see Fig. 3.14).

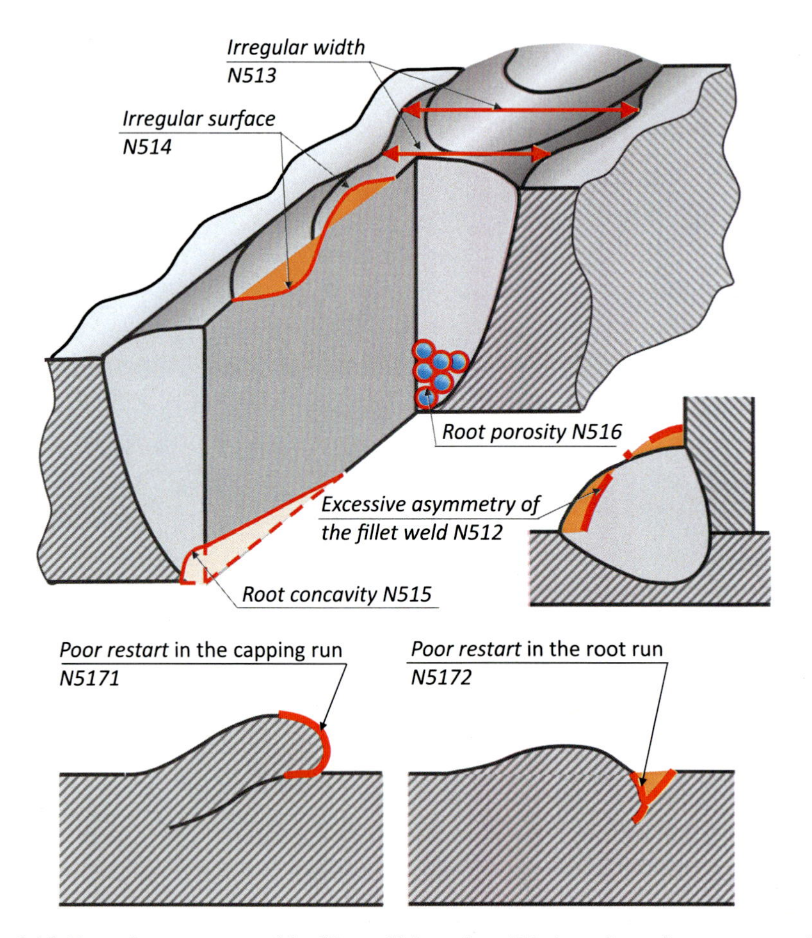

Fig. 3.14 Excessive asymmetry of the fillet weld, irregular width, irregular surface, root concavity, root porosity and poor restart

Irregular width is caused by:

- low qualification of the welder, wrong electrode manipulation,
- long arc,
- unstable welding parameters, improper preparation of edges,
- in MMA welding—use of welding power source with flat volt-ampere characteristic.

(14) **514—irregular surface** (see Fig. 3.14).

Irregular surface is caused by:

- low qualification of the welder,
- uneven welding speed,
- high ductility of the molten flux.

(15) **515—root concavity** (see Fig. 3.14)—shallow deepening in the weld root. Unlike shrinkage grooves (5013) and incomplete root penetration (4021) root concavity is caused by surface tension forces in the lower part of the welding pool and (or) molten flux in case of welding with flux backing.

Root concavity is caused by:

- improper groove shape—big gap, small root face,
- low qualification of the welder,
- high ductility of the molten flux in case of welding with flux backing.

(16) **516—root porosity** (see Fig. 3.14)—sponged zone in the weld root caused by gases freeing up during crystallization.

Root porosity can be prevented using same activities as those for the gas cavities (201_):

- improvement of the welding zone protection, including weld root,
- removal of moisture from welding materials (drying of covered electrodes, fluxes, refinement of shielding gases),
- removal of rust, slag, and oils from edges to be welded and from welding wire prior to welding,
- increase of time when metal is in liquid state, for example by reducing the welding speed,
- increase of welding pool sizes,
- improvement of welder's qualification.

(17) **517_—poor restart** (see Fig. 3.14)—local unevenness on surface in the place of welding renewal. Depending on its location this imperfection may be classified as one of the two types:

- 517**1**—in the capping run,
- 517**2**—in the root run.

Poor restart is caused by:

- low qualification of the welder,

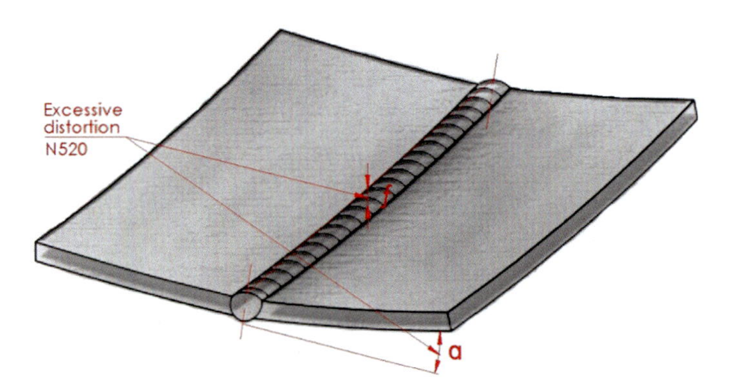

Fig. 3.15 Excessive distortion of the welded joint

- improper welding parameters (low current or voltage).

(18) **520—excessive distortion** (see Fig. 3.15)—deviation of shape and dimensions of the welded joint caused by residual welding stresses and deformations.
Excessive distortion often appears:

- in welded joints of thin plates—due to low stiffness resulting in loss of steadiness under longitudinal and transverse shrinkage forces,
- welded structures made of aluminum and its alloys—as a result of relatively high coefficient of thermal expansion,
- welded structures made of high-alloyed austenitic steels—due to higher coefficient of thermal expansion and lower thermal conductivity in comparison to those of carbon steels.

Excessive distortion can be lowered by preventing its causes, in particular by reducing the residual welding tensions and deformations.

(a) Reduction in heat input:

- welding current kept at a possible minimum,
- increase on welding speed,
- multi-run welding,
- welding without electrode oscillations.

(b) Reduction of zones with elastic–plastic deformations:

- using processes with high energy concentration (laser welding, electron-beam welding, hybrid technologies),
- using fixtures with heat sink,
- copper backings.

(c) Reduction of uneven temperature distribution in the welded joint during welding:

- pre-heating,

- additional heating during welding.

(d) Reduction of length of the zone affected by the shrinkage force where the welded joint loses its steadiness. This is achieved by splitting the longer welds into smaller sections and welding using back-step technique.

(e) Improvement of stiffness of the welded joint:

- structural elements increasing stiffness (ribs, relief, etc.) at the welded structure design stage,
- using jigs and fixtures for fit-up and welding.

(f) Reversal structure deformation during fit-up which compensates residual welding deformations.

(g) Taking into consideration thermal conductivity of the base metal:

- increase of tack welds length and reduction of gaps between them in 1.5–2 times when welding high-alloyed austenitic steels in comparison to welding of carbon and low-alloyed steels,
- cooling to the temperature lower than 100 °C before starting the next run-in case of multi-run welding of high-alloyed austenitic steels.

(19) **521_—imperfect weld dimensions** (see Fig. 3.16)—deviations of thickness and width of the weld from their pre-set values. Imperfect weld dimensions are classified into four types:

- 521**1**—excessive weld thickness. One should take into consideration that weld thickness is a sum of two components: penetration and reinforcement. Both are related to other types of imperfections. Penetration is related to lack of penetration (402) and excessive penetration (504). Reinforcement is related to excess weld metal (502).

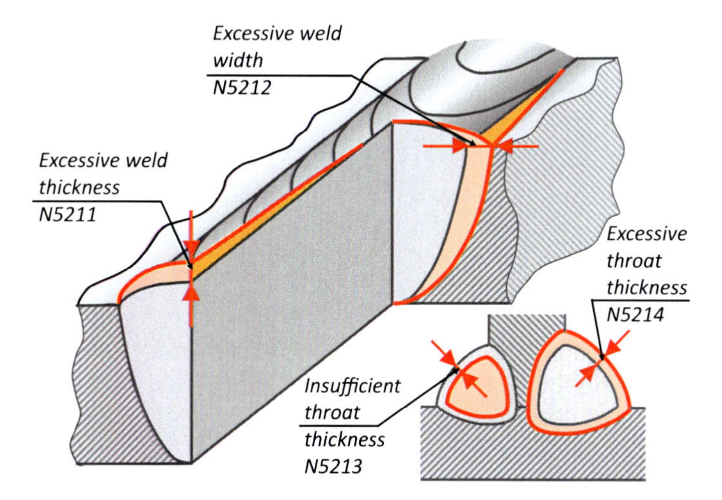

Fig. 3.16 Imperfect weld dimensions

- 521**2**—excessive weld width.
- 521**3**—insufficient throat thickness.
- 521**4**—excessive throat thickness.

Excessive weld thickness is caused by:

- high welding current (main cause), high wire feed speed,
- low welding speed,
- low qualification of the welder,
- big gap between plates to be welded,
- improper groove shape (narrow groove) in case if single-run welding.

Excessive width is caused by:

- high welding voltage (main cause),
- low welding speed,
- low qualification of the welder, improper electrode manipulation.

Lack of throat thickness is caused by:

- low welding current (main cause), low wire feed speed,
- high welding speed,
- low qualification of the welder,
- improper groove shape.

Excessive throat thickness is caused by the same factors as lack of throat thickness (see above):

- high welding current (main cause), high wire feed speed,
- low welding speed,
- low qualification of the welder,
- improper groove shape.

3.1.7 Other Imperfections

(6th group of imperfections)

Group of other imperfections includes 14 subgroups.

(1) **601—arc strike**—local damage of the base metal surface near the weld caused by arc burning outside the groove.
Arc strike is caused by low qualification of the welder.
(2) **602—splatter**—droplets of deposited or filler metal which have fused to the base metal or weld metal surface during welding.
Splatter is caused by:

- low qualification of the welder,

- improper choice of welding materials, in particular shielding gas (MAG welding using gas mixtures reduces splatter, for example 80% Ar + 20% CO_2 mix) and electrode coating type,
- improper dynamic characteristic of welding power source,
- high welding current.

602**1**—**tungsten splatter**—is a variation of splatter listed as a separate imperfection. It appears as parts of tungsten on the surface of the base metal or weld metal.

Tungsten splatter is caused by:

- low qualification of the welder,
- improper shape of tungsten electrode,
- high welding current.

(3) 60**3**—**torn surface**—surface damaged due to removal of temporary auxiliaries.
(4) 60**4**—**grinding mark**—local damage caused by grinding.
(5) 60**5**—**chipping mark**—local damage caused by chisel or other instruments.
(6) 60**6**—**underflushing**—reduction of element thickness due to removal of reinforcement lower than base metal surface.
(7) 60**7**—**tack weld imperfection**—imperfection caused by wrong tack weld placement.

Main cause of torn surface, grinding and chipping marks, underflushing and tack weld imperfections is low qualification of the welder (personnel involved in corresponding operations).

(8) 60**8**—**misalignment of opposite runs** (see Fig. 3.17)—displacement between axis of two runs on the opposite sides of the weld.

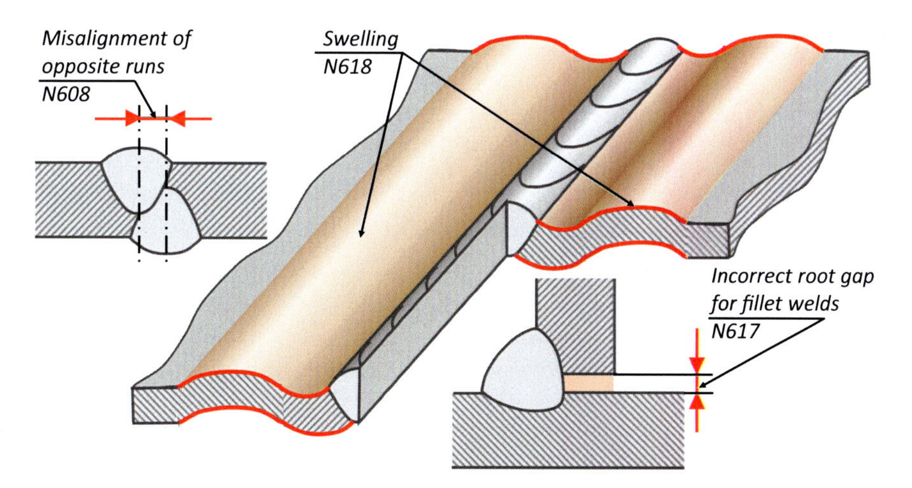

Fig. 3.17 Misalignment of opposite runs

Main cause of misalignment of opposite runs is low qualification of the welder.

(9) **610—temper color** (visible oxide film)—slightly oxidized surface in the welding zone, for example, in case of welding of stainless steel.

Temper colors usually are from yellow to blue due to chromium burn-out in stainless high-alloyed steels and oxidizing of the weld surface.

6101—**discoloration**—is a variation of temper color listed as a separate imperfection. It appears as visible colored surface of deposited metal or HAZ caused by welding heating together with improper gas protection, for example, in case of welding titanium and its alloys.

In welding of titanium dark temper colors (blue and dark blue) indicate significant embrittlement of metal. Lighter colors (light-yellow, yellow, light blue) embrittlement is lower.

(10) **613—scaled surface**—significantly oxidized surface in the welding zone.

Temper color and scaled surface are caused by:

- improper protection of molten and heated metal from the air (main cause),
- improper purity of protective gas (humidity and impurities),
- low qualification of the welder.

Temper color and scaled surface can be prevented by:

- gas protection of the back side of the weld,
- reduction of the heated zone width and temperature (increase of welding speed, using fixtures with heat sink, etc.),
- gas protection after welding is finished until the metal is cooled,
- welding in sealed cameras filled with shielding gas.

(11) **614—flux residue**—flux, not completely removed from the weld surface.

(12) **615—slag residue**—slag, not completely removed from the weld surface.

Flux and slag residues are caused by:

- improper cleaning of weld surface from slag (main cause),
- low qualification of the welder,
- improper flux and electrode cover, that does not provide good removability of slag.

(13) **617—incorrect root gap for fillet welds** (see Fig. 3.17)—excessive or small gap between parts to be welded.

The incorrect root gap for fillet welds is caused by:

- absence or low effectiveness of fixtures,
- low qualification of the welder.

(14) **618—swelling** (see Fig. 3.17)—imperfection caused by long heating of welded joints in light metals during crystallization.

While swelling is caused by excessive heat input, it can be prevented by:

- welding with high energy concentration (laser welding, electron-beam welding, hybrid technologies),
- fixtures with heat sink, for example keyboard clips with directed heat sink.

3.2 Significance of Imperfections. Acceptance Criteria. Testing Levels

Welding as a basic process of manufacturing of welded structures is characterized by complex technique, physical actions and consequences of its effect on metal. Complexity of the process leads to risks of forming of a great variety of defects and imperfections. ISO 6520-1 [1] lists 105 types of imperfections—Sect. 3.1. It is almost impossible to produce a welded joint completely without imperfections.

It is necessary to decide which **imperfections** are **acceptable**—those which do not reduce the rate of conformity of characteristics of welded joint with requirements and expectations of users and other interested parties, that is they do not reduce the quality of welded structure.

This said, **significant defect** is an imperfection which does not correspond to the requirements.

Design perfection together with technical and technological possibilities of manufacturing define a possibility of forming potential imperfections of certain type and in certain amount (Fig. 3.18).

To ensure balance of interested parties' interests (Sect. 3.1.2) it is necessary to achieve a compromise about the significance of imperfections. The compromise is between what manufacturer can provide and what customer is ready to accept. Such compromise leads to decision about acceptable quality level from manufacturer's side on design stage and about acceptance criteria from customer's side on contract development stage.

Quality level is a characteristic of groups of possible imperfections in welded joints defining a list of types of imperfections with their sizes, quantity, and significance.

Standards ISO 5817 [2] (steels, nickel, titanium, and their alloys), ISO 10042 [3] (aluminum and alloys), ISO 13919 series [4, 5] (electron-beam and laser welding) define three quality levels—B, C and D. Standards are applicable for certain welding processes, joint types and material thickness ranges.

Quality level B is high level of quality which allows only some types of imperfections with limited range of dimensions:

- some deviations of form and dimensions—undercuts (5011—not allowed for aluminum and alloys, 5012, 5013), convexity of fillet welds, excess penetration (502, 503, 504), incorrect weld toe (505), linear misalignment (507), incompletely filled groove (511), excessive asymmetry of fillet weld (512), root concavity (515), excessive throat thickness (5214),

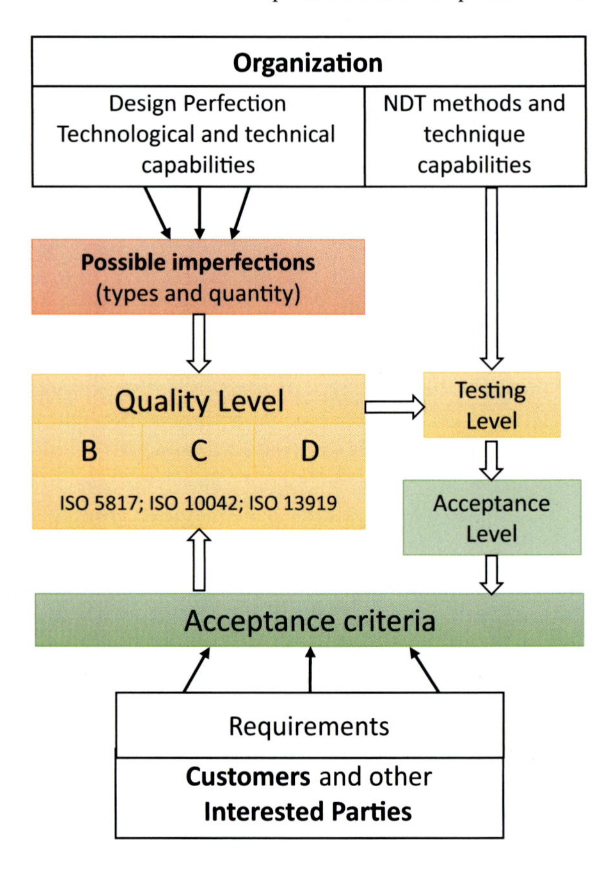

Fig. 3.18 Scheme of meeting the balance of interested parties' interests on significance of imperfections

- microcracks (1001)—significance depends on base metal and its resistance to cracking,
- gas cavities (201)—pores (2011, 2012, 2013, 2014—not allowed for aluminum and alloys, 2015, 2016),
- solid inclusions (300)—slag inclusions (301), flux inclusions (302), oxide inclusions (303), metallic inclusions other than copper (304),
- incorrect root gap for fillet welds (617).

Quality level C—medium quality level (intermediate between B and D).
Quality level D—low quality level which allows most imperfection types except:

- cracks (100),
- lack of fusion (401),
- burn through (510).

The quality levels refer to production quality. They do not refer to the fitness-for-purpose of the product.

A decision on quality level is made taking into consideration the following:

- technological and technical possibilities of manufacturer,
- responsibility of structure which depends on its purpose and risks of failures and breakdowns,
- contract requirements, standards, other relevant regulations,
- mode of stressing (static, dynamic, fatigue),
- service conditions (aggressive environment, high temperatures, etc.),
- subsequent treatment (e.g., coating),
- economic factors (cost of structure production, test, repair, etc.).

Quality level may be assigned:

- same for all welded joints in the structure,
- different for different welded joint in the structure,
- different for different imperfections in the same welded joint.

For example, when Chornobyl nuclear power plant confinement was designed, all welds were assigned to the highest quality level B. Exception was made for welded joints on lugs which were assigned quality level D. This decision was made because lugs were to be used only during confinement fit-up and in-service risks related to their failure are insignificant.

Customer's requirements are defined in acceptance criteria. They include significance of imperfections which customers agree to accept.

Acceptance criteria are characteristics of welded structure with deviation tolerances including types, size and quantity of insignificant imperfections which satisfy the customer. Imperfections are to be checked by manufacturer with final testing and/or by the customer with initial one. Acceptance criteria are usually listed in product specification and/or in the contract.

Acceptable quality level may be assigned as acceptance criterion. Numbering of criteria may be the same as that for quality levels or may differ, e.g., like in the ISO 17635 [6].

Requirements to imperfection characteristics on different stages of welding manufacturing are defined by standards (see Table 3.1).

To control quality level and meet acceptance criteria it is necessary to choose method and technique of non-destructive testing which ensure sensitivity to imperfections sought and can define parameters of imperfections relevant to the quality level. This is done by assigning the testing level.

Table 3.1 Imperfections requirements standards

Stage of welding manufacturing	Preparation cutting	Welding including post-welding treatment	Subsequent treatment preparation to coating	Acceptance NDT
Standards on imperfections	ISO 9013	ISO 5817, ISO 10042, ISO 13919	ISO/TR 15,235	ISO 17635

Testing level—test volume (combination of testing methods, test area, scanning technique, requirements to other characteristics of testing technique, requirements to documented procedure, etc.) and values of imperfections' parameters on which testing is performed. Testing level is a characteristic of possibilities of testing method and technique as well as that of possibility of detection of imperfection.

Testing level is assigned for a particular welded structure testing method based on the following:

(a) Requirements of international standards covering the testing method.
(b) Assigned quality level of welded structure.
(c) Technological factors:

- welding technology,
- base metal and welding materials,
- type and size of joints,
- configuration of parts (accessibility, state of surface, etc.),
- type of possible imperfections and their orientation.

Testing on the assigned level allows to obtain the corresponding quality level. Tables of correspondence between testing levels and quality levels are given in ISO 17635 [6] for basic NDT methods as well as in standards covering each NDT method.

Acceptance level is a criterion of testing method. If this level is exceeded corrective actions are to be performed. Acceptance level ensures correspondence to acceptance criteria for significance of imperfections.

This said, quality level, acceptance criteria, testing level and acceptance level are instruments for achieving the balance of interested parties' interests in solving complex and responsible problem of significance of imperfections in welding manufacturing.

3.3 Engineering Critical Assessment Techniques

3.3.1 General

When designing welded structures, a decision can be made on the imperfection admissibility in any welded joint, regardless of the quality level adopted for the entire structure (Sect. 3.2). In this case, the type and size of permissible imperfections must be determined. The main requirement for a welded structure is to ensure the strength of welded joints. The strength of a welded joint with an imperfection is determined by the stress–strain state in the imperfection zone. To characterize the stress–strain state in the imperfection zone and determine the critical imperfection size in engineering calculations, two criteria are used:

(1) Stress concentration factor (Sect. 3.3.2) is used for:

- Cavity (2nd imperfection group per ISO 6520-1 [1]), excluding interdendritic shrinkage (2021),
- Inclusions (3rd imperfection group), excluding oxide inclusions (303),
- Incomplete penetration (402),
- Imperfect shape and size (5th imperfection group).

(2) K_I stress intensity factor (Sect. 3.3.3) is used for:

- cracks (1st imperfection group per ISO 6520-1 [1]),
- crack-like imperfections—interdendritic shrinkage (2021), oxide inclusions (303) and lack of fusion (401).

3.4 Determination of Critical Defect Size by Stress Concentration Factor

When determining critical imperfection size by stress concentration factor, first failure theory of applied mechanics is applied, according to which failure occurs when normal stress of a critical level (ultimate strength limit) is reached.

Permissible stress [σ] is the highest stress at which trouble-free operation of the structure is ensured.

Permissible stresses for a given type of the welded structure are determined, as a rule, by dividing critical stress level (σ_c) by load factors (n_i):

$$[\sigma] = \sigma_c / n_i \tag{3.2}$$

The denominator of the Eq. (3.1) can have one load factor or a product of several load factors, which depend on:

- metal state (brittle or ductile) and the nature of the operating loads (static, dynamic, fatigue),
- structure liability, which is determined by its purpose, and the risks of failures and incidents,
- operating conditions (aggressive environment, temperature),
- inaccuracies in setting external loads and approximation of design schemes.

The strength condition is when the maximum stresses (σ_{max}) in the most critical area do not exceed the permissible stress values:

$$\sigma_{max} \leq [\sigma] \tag{3.3}$$

For welded joints, sections with imperfections are the critical area.

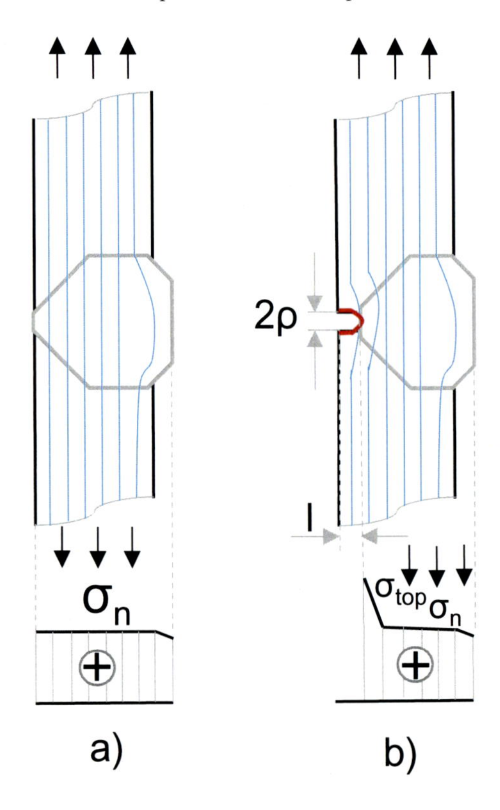

Fig. 3.19 Force filed lines and stress distribution diagrams in welded joint cross-sections: **a** without imperfection, **b** with imperfection

When an imperfection-free welded joint is stress strained, the force field lines are evenly distributed over the cross section, the stress diagram has a rectangular shape (Fig. 3.19a). A slight decrease in stresses in the excess weld metal is usually not taken into account in engineering calculations.

For a stress strained welded joint with an imperfection, for example, lack of fusion in the root, the metal above the imperfection does not accept the load. Therefore, the metal at the top of the imperfection must accept an increased load. The density of the force field lines at the top of the imperfection increases—the force field is concentrated (Fig. 3.19b). In this case, in the stress distribution diagram an extremum appears at the top of the imperfection. Imperfection is a stress concentrator.

Stress concentrator—a sharp local change in cross-sectional area, leading to an uneven distribution of stresses.

A quantitative characteristic of a stress concentrator effect on the stress level is the **stress concentration factor**—the ratio of the highest local stress at the top of the concentrator (σ_{top}) to the nominal stress (σ_n) calculated based on the working cross section assuming the absence of the concentrator:

$$\alpha = \sigma_{top}/\sigma_n \tag{3.4}$$

Stress concentration factor depends on:

- imperfection shape,
- imperfection size,
- imperfection top radius,
- stress type (strain, bend, shear, etc.) at the cross-section.

This is the condition for the strength of a welded joint with an imperfection taking into account stress concentration:

$$\alpha\, \sigma_{\max} \leq [\sigma] \tag{3.5}$$

To determine critical size of an imperfection by stress concentration factor:

1) Determine permissible stresses [σ] for the given welded joint (3.2). Safety factors are determined from regulatory documents (industry standards, specifications, etc.). If there are no regulatory documents, safety factor can be chosen from the values obtained experimentally:

   - $n_i = 1$—corresponds to 50% failure probability,
   - $n_i = 1{,}2$—corresponds to 90% failure probability,
   - $n_i = 1{,}5$—corresponds to 99% failure probability,
   - $n_i = 2$—corresponds to 99.9% failure probability.

(2) Calculate maximum workload stress σ_{max} in section without imperfection.

(3) Select suitable stress concentrator form from the reference, as close as possible to the shape of the imperfection. The stress concentration factor for the selected concentrator form will be a:

   - mathematical dependence on the size of the concentrator,
   - constant.

(4) Substitute stress concentration factor value in the strength condition (3.5) and determine:

   - critical dimensions of the concentrator (defect), if the stress concentration factor is a mathematical dependence on the dimensions of the concentrator,
   - maximum permissible stresses σ_{max} in the cross section, taking into account the imperfection, if stress concentration factor is a constant.

Example 1 Determination of critical level of linear misalignment between plates (imperfection 5071—Fig. 3.11).

Stress concentration factor for the selected concentrator shape:

$$\alpha = 1 + 3\Delta/\delta \tag{3.6}$$

Δ—Level of linear misalignment between plates.

Δ—plate thickness.

Strength condition of a welded joint with imperfection:

$$(1 + 3\Delta/\delta)\,\sigma_{\max} \leq [\sigma] \tag{3.7}$$

Critical level Δ_C linear misalignment between plates:

$$\Delta_C = ([\sigma]\,/\,\sigma_{\max} - 1)\,\delta\,/3 \tag{3.8}$$

Example 2. Determination of critical depth of incomplete root penetration (imperfection 4021)—Fig. 3.19b.

In relation to incomplete root penetration K. Inglis solution for undercut is used (assumption is made that incomplete penetration top has semi-circular shape):

$$\alpha = 1 + 2\sqrt{(l/\rho)} \tag{3.9}$$

l—undercut depth (incomplete penetration),

ρ—radius of curvature at the top of the cut.

Strength condition of a welded joint with imperfection

$$\left(1 + 2\sqrt{(l/\rho)}\right)\sigma_{\max} \leq [\sigma] \tag{3.10}$$

Critical depth of incomplete penetration l_C with the gap 2ρ between welded edges:

$$l_C = ([\sigma]\,/\,\sigma_{\max} - 1)^2\rho\,/4 \tag{3.11}$$

3.5 Determination of Critical Length of Cracks and Crack-Like Imperfections

Cracks are stress concentrators with a tip radius ρ = 0. In this case, the stress concentration factor becomes equal to infinity. Assessment of fracture by the critical stress level at the crack tip loses its meaning.

In linear fracture mechanics, the distribution of stresses and strains in the region adjacent to the crack tip is used as a criterion characterizing the possibility of crack development (the beginning of fracture). The nature of this distribution is described by the K_I stress intensity factor.

Stress intensity factor K_I—function, that characterizes the force field in the crack tip area and is a quantitative measure of stresses and strains at the tip.

Stress intensity factor depends on:

- nature of object in which the crack propagates (infinite or semi-infinite plate, strip, etc.).
- location (internal, surface), shape and orientation of the crack.
- nature of the load (tensile, transverse misalignment, longitudinal misalignment, bending).

- active stresses.
- other parameters (distribution of temperature, etc.).

The condition for the crack growth start is when stress intensity factor (K_I) is superior to critical value (K_{IC}):

$$\kappa_I \geq \kappa_{IC} \tag{3.12}$$

Stress intensity critical factor K_{IC} (or crack resistance)—mechanical characteristic of the material, which defines material resistance to crack propagation and is determined experimentally (similar to tensile strength, impact strength, etc.)

Condition (3.12) is called the strength criterion of G. Irwin.

To determine critical length of cracks and crack-like imperfections by stress intensity factor:

(1) From reference sources, select a formula for calculation of stress intensity factor of the crack, which is most suitable for the nature of the object in which the crack propagates, location, shape and orientation of the crack, as well as the nature of the load.
(2) Calculate or determine by non-destructive testing methods (Chap. 4) the level of current stresses, which are included in the formula for calculating the stress intensity factor of a crack.
(3) Determine from critical stress intensity factor reference sources or experimentally.
(4) From the condition for the crack growth start (3.12) determine the critical length of the crack.

Example 1. Determination of critical length of a surface semi-elliptical crack

The formula for calculating stress intensity factor of a surface semi-elliptic crack located in the normal stress field σ is:

$$\kappa \mathrm{I} = \sigma \sqrt{(\pi \mathrm{a})} \tag{3.13}$$

a—half-length of the major axis of a semi-elliptic crack

After substituting (3.13) into (3.14), we obtain the critical value of the half-length of the surface semi-elliptic crack:

$$\mathrm{a_c} = \left(\kappa_{\mathrm{IC}}/\sigma^2\right) 1/\pi \tag{3.14}$$

Literature References

1. ISO 6520-1:2007, Welding and allied processes—classification of geometric imperfections in metallic materials—Part 1: Fusion welding

2. ISO 5817:2014, Welding—fusion-welded joints in steel, nickel, titanium and their alloys (beam welding excluded)—Quality levels for imperfections
3. ISO 10042:2018, Welding—arc-welded joints in aluminum and its alloys—quality levels for imperfections
4. ISO 13919-1:2019, Electron and laser-beam welded joints—requirements and recommendations on quality levels for imperfections—Part 1: Steel, nickel, titanium, and their alloys
5. ISO 13919-2:2021, Electron and laser-beam welded joints—requirements and recommendations on quality levels for imperfections—Part 2: Aluminum, magnesium and their alloys and pure copper
6. ISO 17635:2016, Non-destructive testing of welds—general rules for metallic materials

Chapter 4
Non-destructive Testing

4.1 Nondestructive Testing Objects and Selection of Methods

4.1.1 Nondestructive Testing Objects

Nondestructive testing (NDT)—is a group of methods that allow to perform testing directly on the structure without its destruction to prepare samples, drilling, etc.

Main purpose of NDT of welded joints is to evaluate the rate of changes in service characteristics of welding resulting from physical, chemical, and metallurgical effects. Physical, chemical, and metallurgical effects of welding include many factors and differs not only for welding methods, but for structures of the same type as well.

Physical, chemical, and metallurgical effects of welding is defined by two groups of factors (see Fig. 4.1):

- technological factors and disturbances related to them (uncontrolled effects which deviate process of welding from its normal way—see Sect. 1.5.4.2),
- environmental factors.

(1) Technological factors are:

 (a) Heating from different sources. Energy sources input energy into the welding zone. For most welding processes this energy is much higher than that required to form a welded joint. Excess of energy leads to an increase of maximum temperatures reached as well as to increase and changes in thermal cycles in the heat-affected zone. This results in the forming of different structures in different zones of the welded joint. Significant unevenness of heating and cooling in different zones of the welded joint leads to residual welding stresses and deformations. For fusion welding their values in the weld are comparable with that of yield strength. Overheating of the welding zone adds residual stresses and deformations create

S. Fomichov et al., *Quality Management in Welded Fabrication*,
https://doi.org/10.1007/978-3-031-34800-6_4

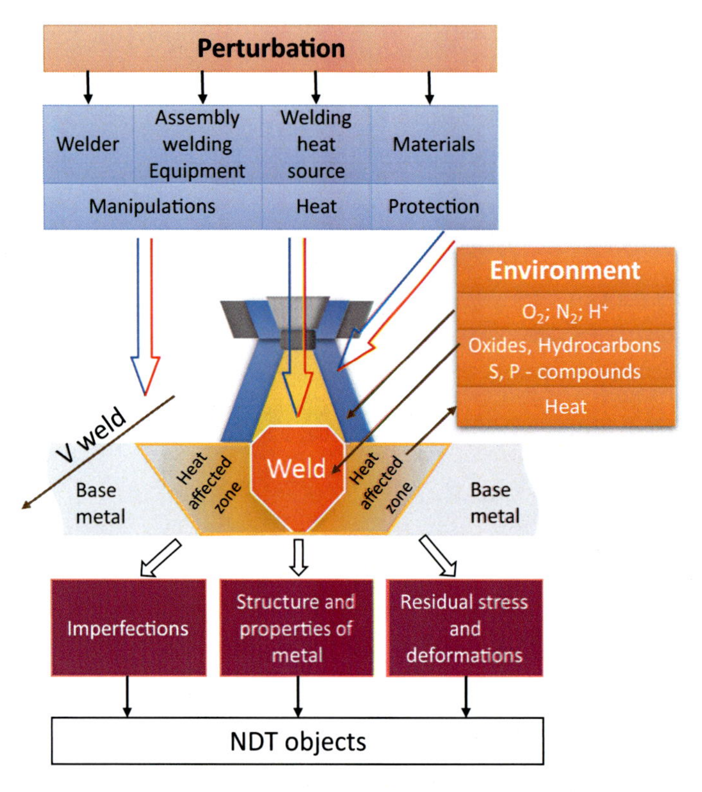

Fig. 4.1 Formation of NDT objects

conditions for formation of hot and cold cracks (100, Sect. 3.1), burn-through (510, Sect. 3.1), undercuts (501, Sect. 3.1), angular misalignment (508, Sect. 3.1), distortion (520, Sect. 3.1), swelling (618, Sect. 3.1).

(b) Manipulations performed by assembly and welding equipment and by the welder. Precision of structure position and movement of the welding heating source affect possibility of formation of shrinkage cavities (202, Sect. 3.1), lack of fusion (401, Sect. 3.1), majority of improper shape imperfections (5th group, Sect. 3.1), arc strikes, splatter, torn surface, grinding and chipping marks, underflushing and tack weld imperfections (601–607, Sect. 3.1), misalignment of opposite runs (608, Sect. 3.1), incorrect root gaps in fillet welds (617, Sect. 3.1)

(c) Protection of the welding zone with materials (gases, fluxes) is ensured by equipment. Correctness of choice and cleanliness of the base metal, welding wire, gases, and fluxes together with effectiveness of gas nozzles and other units related to the welding zone protection create conditions for formation of the following imperfections:

- gas cavities (201, Sect. 3.1),

- slag (301, Sect. 3.1), flux (302, Sect. 3.1) and oxide (303, Sect. 3.1) inclusions,
- lack of fusion (401, Sect. 3.1),
- undercuts (501, Sect. 3.1),
- incorrect weld toe (505, Sect. 3.1),
- overlap (506, Sect. 3.1),
- root porosity (516, Sect. 3.1),
- splatter (602, Sect. 3.1),
- temper color (610, Sect. 3.1),
- scaled surface (613, Sect. 3.1)

Environmental factors affecting welding are:

(a) Effect of gases (O_2, N_2, H^+) from the atmospheric air leads to formation of gas cavities (201, Sect. 3.1), as well as hydrogen-induced cracks (100, Sect. 3.1).
(b) Impurities in the welding zone—oxides, hydrocarbons, compounds containing sulfur and phosphorus. Impurities similarly to the effect of gases from the air are related to poor protection and lead to imperfections listed in 2c.
(c) Uncontrolled heat sink leads to uneven cooling of different zones of the welded joint. This increases effect from excessive energy input. Uncontrolled heat sink is one of main causes of structure nonuniformity between zones of the joint, welding stresses and deformations as well as that of imperfections listed in 1a.

This said, the results of physical, chemical, and metallurgical effects of welding which led to changes in service characteristics of the welded structure are:

(1) Imperfections,
(2) structure changes in the joint,
(3) residual stresses and deformations.

It should be mentioned that most NDT methods are aimed at detecting imperfections in the welded joints and surrounding zones.

4.1.2 Selection of NDT Methods Versus Applications

Welded structures undergo non-destructive testing during their whole life cycle, starting from manufacturing itself. The applicable NDT methods are plenty, the standard for NDT personnel qualification and certification lists 10 of them directly with possible extension of standard recommendations to other NDT techniques.

Each method has its own advantages and disadvantages. Generally speaking, application of any NDT method highly depends on the geometrical conditions of the component to be tested as well as on the configuration and accessibility of the joint. This is particularly true for volumetric methods, such as RT and UT.

The methods for surface testing (VT, MT, PT, and Eddy current) are primarily dependent on the surface conditions and accessibility. It should be mentioned that surface state remains quite important for all methods which imply using sensors as far as a poor contact between the testing object and the sensor itself may affect the defect detection.

In addition, NDT methods are limited by physical phenomena used in each technique, equipment (type, portability, cost, etc.), applicability for a certain material or structure.

For example, in UT—the direction of the US-beam, and in RT—the direction of the radiation beam relative to the anticipated imperfections, are very important. Also, the metallurgical structure of the material and its thickness can have a significant effect on the NDT method effectiveness.

Testing conditions are important as well: where, when, how frequently and for how long the testing should be performed.

Technical Report CEN/TR 15135 [1] is an informational document. It is aimed at helping with design and evaluation of various joint types and geometrical configurations. This is a guidance for evaluating the accessibility of the weld for NDT in general or its ability to be examined with a particular NDT method. Examples given in the document will help with planning for NDT during design and fabrication.

In CEN/TR 15135 [1] the general recommendations for five NDT methods are given: VT, UT, RT, MT, and PT. These recommendations can easily be extended to other methods using similar techniques or based on similar physical phenomena.

Each method is described by two characteristics:

- whether it is applicable at all, if yes—are there any limitations,
- whether results satisfy ordinary requirements, if no—can they be improved by combining with other NDT method(s).

The document also takes into consideration evaluation of design for NDT (e.g., the particular method presumably is highly effective for detection of imperfections in the weld, but the weld itself is inaccessible for this method). The range of geometrical conditions in welded structures is very wide and have various applications. While it is practically impossible to provide a clear unequivocal solution for every case alone, the typical designs are listed in the CEN/TR 15135 [1]. Each of them includes several joint configurations, divided into groups:

- the least acceptable configuration for testing,
- better joint configuration for testing,
- the best configuration,
- additional modifications.

While documents similar to CEN/TR 15135 [1] provide recommendations, it is important to remember that NDT method choice as well as structure design and planning of manufacturing, operating, service and in-service inspection should also be based on sectoral requirements (e.g., NDT method can be demanded by a regulation or code) and resources of organizations, involved on different stages of the structure's life-cycle.

4.2 Organoleptic Methods. Visual Testing

4.2.1 Fundamentals of Organoleptic Methods

Organoleptic methods are based on using sense organs for expert evaluation of quality characteristics of products and services:

- vision (visual testing)—to determine presence of something (surface imperfections), color, shape, approximate dimensions.
- Smell—to determine presence of toxic compounds in the air (for example, smell of smoke can be a basis for starting fire-fighting activities), search for narcotics and explosives (with use of trained dogs).
- Touch—to determine tactile properties, approximately roughness, toughness, hardness.
- Taste—for degustation of drinks and food.
- Hearing—for evaluation of sound characteristics of acoustic systems, for definition of noises.

Visual testing (VT) is widely used in welding manufacturing.

4.2.2 Application Area of Visual Testing

VT is performed during any control as a separate method or as a first one in combination with other methods.

As a separate method VT is used to define:

- deviations of shape and size (5th group of imperfections—Sect. 3.1),
- lack of fusion in the butt weld root (4021—Sect. 3.1),
- other imperfections (6th group of imperfections—Sect. 3.1).

VT combined with other methods is used to evaluate:

- all types of cracks (1st group of imperfections—Sect. 3.1),
- cavities (2nd group of imperfections—Sect. 3.1).

According to ISO 17635 [2] VT can be used for all types of surface imperfections both in butt and fillet welds, for a variety of metals:

- ferritic steels,
- austenitic steels,
- aluminum and alloys,
- nickel and copper alloys,
- titanium and alloys.

4.2.3 VT Technique

VT technique of joints welded with fusion welding methods is described in ISO 17637 [3].

VT should be performed under the surface luminosity higher than 300 lx, under the vision angle more than 30 °C, at a distance approximately 600 mm (see Fig. 4.2).

During VT additional technical means can be used, in particular:

- mirrors, optic cable, or camera—for remote testing,
- additional sources of light—for improvement of luminosity, contrast, and clarity,
- magnifying glasses and other means of image zoom and magnification—for small sized imperfections,
- ruler, calipers, templates, and other measuring instruments.

Geometric dimensions of welded joints are often determined with special templates. Figure 4.3 shows a sample of universal welder's template.

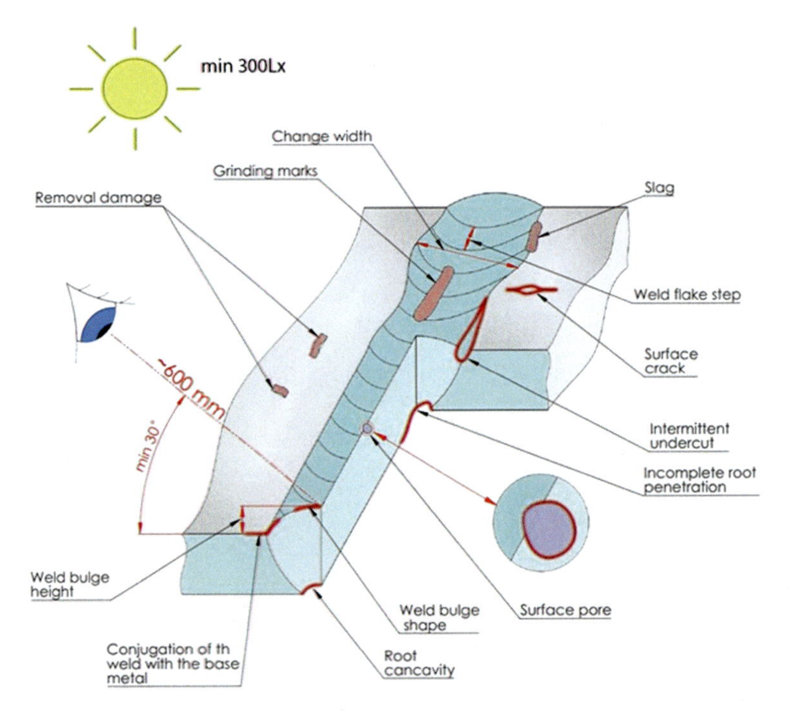

Fig. 4.2 Visual testing. main objects

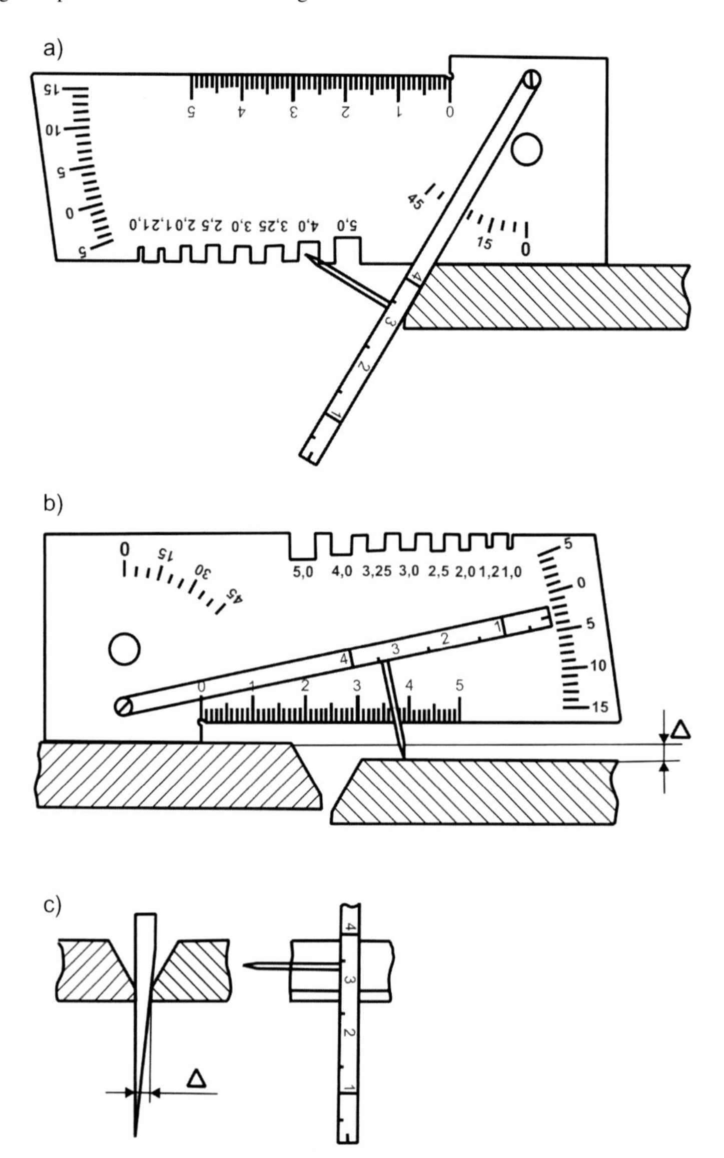

Fig. 4.3 Diagrams of application of the universal welder's template: **a** control of the bevel angle, **b** measurement of linear displacement of the surfaces of the assembled parts, **c** measurement of the gap between the assembled parts, **d** measurement of the bulge of the butt weld, **e** measurement of the width of the weld

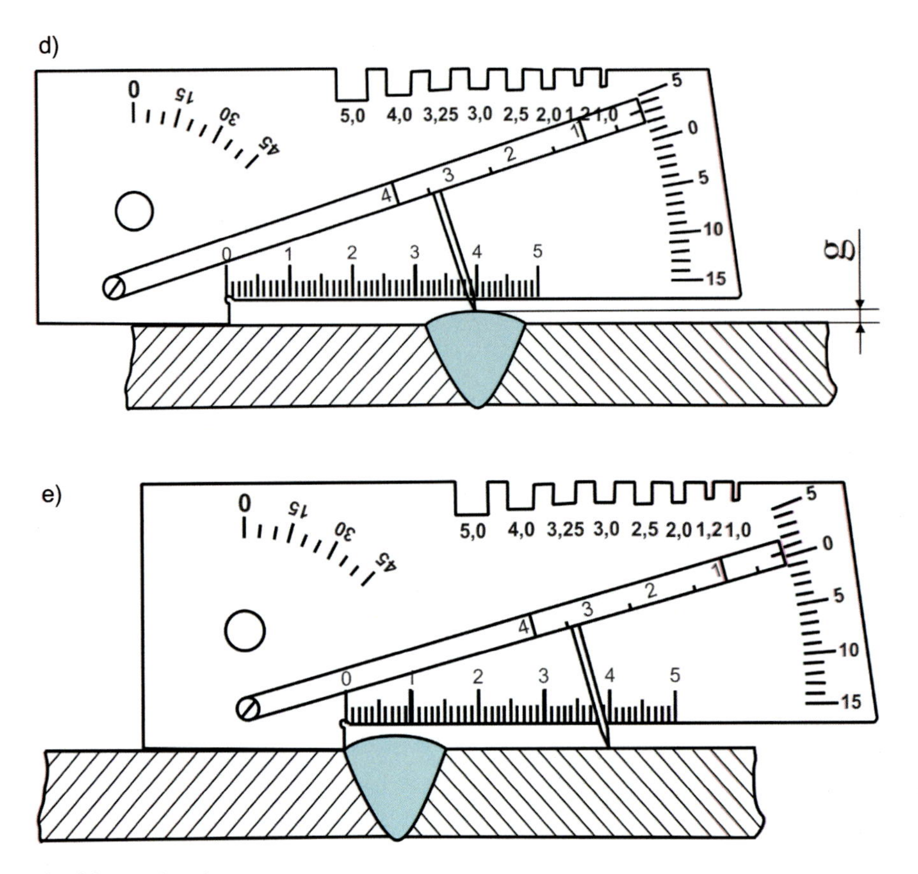

Fig. 4.3 (continued)

During VT it is necessary to define if characteristics of the welded joint correspond to those listed in the Welding Procedure Specification (WPS), product standards and acceptance criteria. VT can be performed before, during and after welding.

(1) Before welding VT is used to control:

- shape and dimensions of groove,
- cleanliness and state of surface of the edges to be welded,
- proper fixation of parts to be welded.

(2) During welding VT is used to control:

- cleaning of each bead of the weld before starting the next run,
- absence of visible imperfections, e.g., cracks and cavities,
- fullness of transition between the weld runs and base metal for good fusion during the next run.

(3) After welding VT is used to control:

- cleanliness and treatment of the weld (removal of slag, absence of chisel and grinding marks, smoothness of transition from the weld to the base metal),
- shape and dimensions of the weld (reinforcement shape, stability of width along the weld, look of the weld surface, including pattern and step of scales),
- weld root and face (fullness of penetration, root concavity, absence of burn-through, shrinkage and undercuts along the weld, absence of significant imperfections on the surface, such as cracks and pores, absence of marks after removal of temporarily welded auxiliary parts).

(4) In case of repairs VT is used to control:

- proper shape and depth of the groove after removal of the original weld,
- correspondence of the repaired weld to requirements listed for the original one.

4.2.4 Advantages and Limitations of VT

VT advantages are:

(1) Informational content—results of VT give information about presence, type, dimensions, and location of surface imperfections.
(2) Universality—method can be used for any type of welded joints produced with fusion welding methods in structures made of any material.
(3) Simplicity and low cost—method does not demand using of complex and costly equipment, only simple instruments.

VT limitations are:

(1) Method cannot be used to define interior imperfections.
(2) VT results significantly depend on personnel competency. Desired competency is defined based on responsibility of welded structures. It is recommended that personnel are qualified according to ISO 9712 [4] or similar standard for the manufacturing area.

4.3 Ultrasonic Testing

4.3.1 Method Fundamentals

Ultrasonic testing (UT) is one of the most widely used NDT methods (combined with VT). It is used to detect:

(a) presence of discontinuity-type imperfections—cracks, cavities, solid inclusions, lack of fusion and penetration (imperfection groups 1, 2, 3 and 4—Sect. 3.1), as well as root porosity (516—Sect. 3.1),
(b) defect location,
(c) defect dimensions,
(d) quantity of imperfections.

UT is based on three physical phenomena:

(1) **Inverse Piezoelectric Effect**—mechanical deformations of piezoelectric plate under effect of electric field. If alternating voltage with ultrasonic (US) frequency is applied to piezoelectric plate surface, the plate starts vibrating at US frequency. In this way pulsed mechanical oscillations with US frequency (ultrasonic beam) are generated in the metal.
(2) **US-wave reflection from interface (defect surface)**. Due to this effect detection of welding imperfections is possible.
(3) **Direct piezoelectric effect**—when piezoelectric plate is being deformed, electric charges with different signs are generated on its surface. Due to this the defect echo is transformed into US electric signal by transducer for further amplification and indication.

During UT, the primary US transducer works in two modes (see Fig. 4.4):

- as a pulse generator—during short time,
- as an echo receiver—during longer time.

In generator mode (blue lines on Fig. 4.4) two electric pulses are sent synchronically:

- Transmission pulse indication—on the display of defect detector. Transmission pulse indication is a beginning of coordinates to define echo time.
- Transmitter pulse—to the surface of piezoelectric primary transducer. As a result, ultrasonic beam is generated in the metal.

To apply US beam to the metal and to ensure imperfection echo returning to the control zone:

- the surface should have a particular roughness (for example, achieved by grinding), because microroughness scatter the US wave,

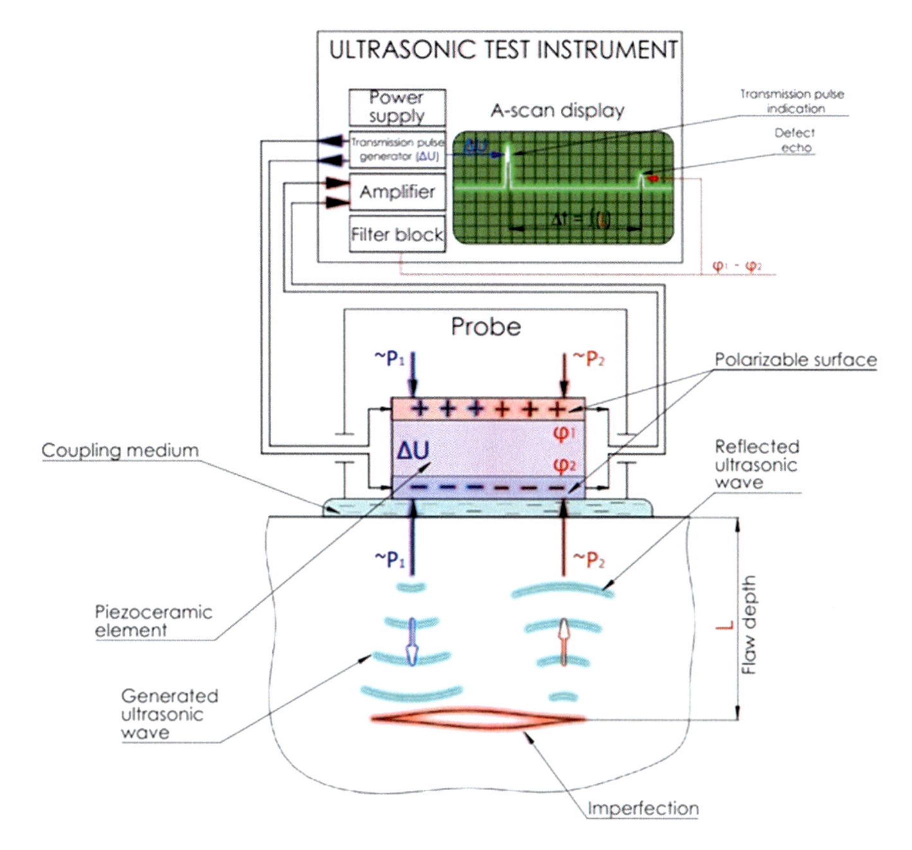

Fig. 4.4 UT method scheme

- coupling medium should be applied between transducer and testing object, because liquid interface reduces coefficient of reflection in the US wave probe index.

If there is a defect on the way of the US beam, US wave reflects from its surface and defect echo is generated.

In the receiver mode (red lines on Fig. 4.4) defect echo initiates mechanical oscillations of piezoelectric transducer. Electric potentials ϕ_1 and ϕ_2 appear on transducer's surface. Potential difference $\mathrm{E} = \phi_1 - \phi_2$ is proportional to the defect echo. Electric pulse is indicated on the display of defect detector. Testing results are shown as a graph with X-line corresponding to time and Y-line—amplitude of US-signal. This is called A-scan display.

Defect's coordinate X (distance between the place of pulse input and defect's surface by which the imperfection depth is evaluated) is defined by time Δt between transmission pulse indication and defect echo:

$$X = v\Delta t/2 \tag{4.1}$$

v *sound velocity in metal*

Other two coordinates (Y, Z) are defined by location of US-transducer at the time when the imperfection was detected.

During the period (T), when US-transducer works in the generator mode it cannot receive defect echoes. Because of this the so-called "dead zone" is created. "Dead zone" depth (H):

$$H = v\,T/2 \tag{4.2}$$

At the depth of the "dead zone" defects cannot be detected by the direct US beam (see below).

The smaller the imperfection, the lesser the amplitude of defect echo. Amplitude characterizes the area of reflection, by which size of defect can be evaluated. In addition, dimensions of continuous defect can be determined by moving US-transducer by the distance on which the defect echo is detected.

Defect detection sensitivity—ability of US defect detector to identify the minimal defect echo.

Defect detention sensitivity is set during US defect detector adjustment prior to testing. If sensitivity is increased the 'grass' starts appearing on the screen—multiple chaotic echoes induced by re-reflection of US waves from grain borders and other structural heterogeneities which, in fact, are not imperfections.

To ensure unification between set-up and results' presentation before testing a reference level is set as a scale of the Y-axis of US defect detector. To set the reference level special reference blocks are used (for example, with holes 3 mm diameter) to set the required amplitude value.

During testing indications of amplitude on US defect detector are compared to the evaluation level.

Evaluation level—an indication on US defect detector. If the signal amplitude exceeds evaluation level it is necessary to analyze the corresponding imperfection and to perform corrective actions when they are applicable (Sect. 1.5.13).

Time base range—minimum and maximum time values (or distance values) within which signals are indicated on the screen. In other words, this is a scale of X-axis of the US defect detector.

Volume of control—zone of the welded joint which must be tested. According to requirements of international standards on UT (see below) the volume of control should include the weld itself and 10 mm of the base metal on each side of the weld (or the whole width of HAZ, whichever is bigger).

Terms and definitions related to US testing are listed in ISO 5577 [5].

According to ISO 17635 [6] UT can be applied to detect internal defects of welded joints in a variety of metals:

- ferritic steels—butt and fillet welds without limitations in thickness,
- austenitic steels—butt welds (thickness more than 8 mm) and fillet welds without limitations of thickness,

- aluminum and alloys—butt welds (thickness more than 8 mm) and fillet welds without limitations of thickness,
- nickel, copper, and their alloys—butt welds (thickness more than 8 mm) and fillet welds without limitations of thickness,
- titanium and alloys—butt welds (thickness 8 to 40 mm) and fillet welds (thickness up to 40 mm).

Depending on the number of primary transducers and control of US pulses input and echo processing there are three UT techniques:

(1) Ultrasonic pulsed echo technique (UT-PE).
(2) Time-offlight diffraction technique (UT-TOFD).
(3) Phased array ultrasonic technique (PAUT).

4.3.2 Ultrasonic Pulsed Echo Technique (UT-PE)

UT-PE—is a traditional ultrasonic technique with the following features:

(1) US probe includes one piezoelectric element (rarely—two, when one works in generator mode and the second one in receiver mode).
(2) **Defect echo** is informative parameter.
(3) Information is displayed as A-scan display (see above Sect. 4.3.1.

Two types of US probes are used in UT-PE technique (see Fig. 4.5): normal probe and angle probe.

In the normal probe the ultrasonic beam input is perpendicular to the product surface. With this technique it is possible to detect only those imperfections which are located under the probe. Weld reinforcement makes it impossible to place the probe upon the weld, so the direct scan technique can be used only to detect imperfections in the HAZ or to define the part thickness.

The angle probes are used in the following angle beam techniques:

- direct scan technique—to detect imperfections in the HAZ and in the weld root,
- double traverse technique—to detect imperfections on the whole thickness of the weld including those located in the "dead zone",
- multiple traverses' technique—to detect imperfections in metal with small thickness (up to 8 mm). Sensitivity of multiple traverses is reduced because due to multiple reflection part of US-pulse energy dissipates.

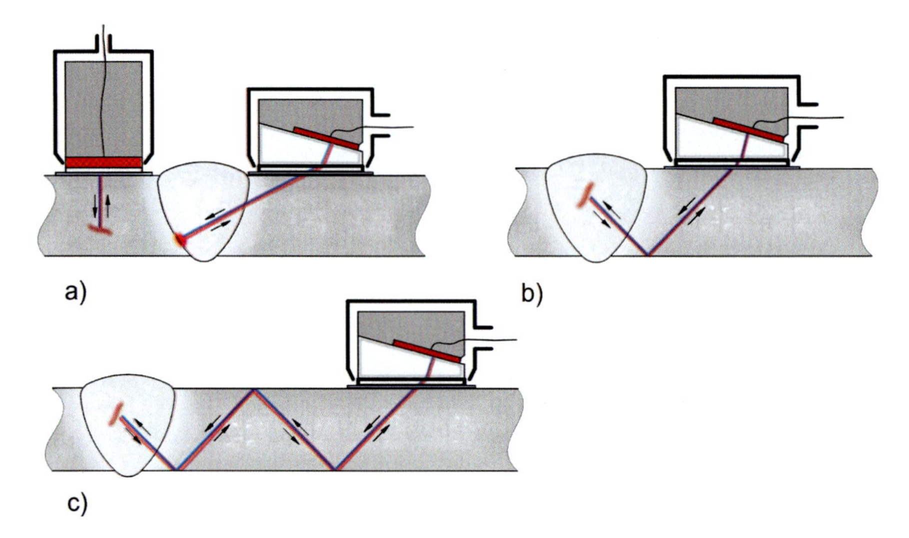

Fig. 4.5 Types of US probes and techniques of testing: **a** direct scan technique, **b** double traverse technique, **c** multiple traverse technique

4.3.3 Time-Offlight Diffraction Technique (UT-TOFD)

UT-TOFD—is the ultrasonic technique with the following features:

(1) Physical phenomenon of diffraction of waves is used—when waves scatter around the obstacle (defect). Defect edge becomes the secondary waves generator. Interference of waves takes place outside the imperfection edge (see below Sect. 4.3.4).
(2) **Two angle probes** located on different sides of the weld are used for testing. The first probe generates the pulse which creates an interference pattern outside of the imperfection. The second probe receives US waves scattered from the defect edges.

The main informative characteristic in UT-TOFD is a difference in time between four pulses on the screen (see Fig. 4.6):

- LW—from the lateral wave generated by the first probe,
- UT, LT—from diffracted waves appearing on imperfection edges in its upper and lower part respectively,
- BW—from the back-wall echo generated by the first probe.

Size of imperfection is defined by the difference in time between the second angle probe is reached by diffracted waves appearing on the imperfection edges (distance between UT and LT pulses). Pulse amplitude is not used for estimation of the imperfection size.

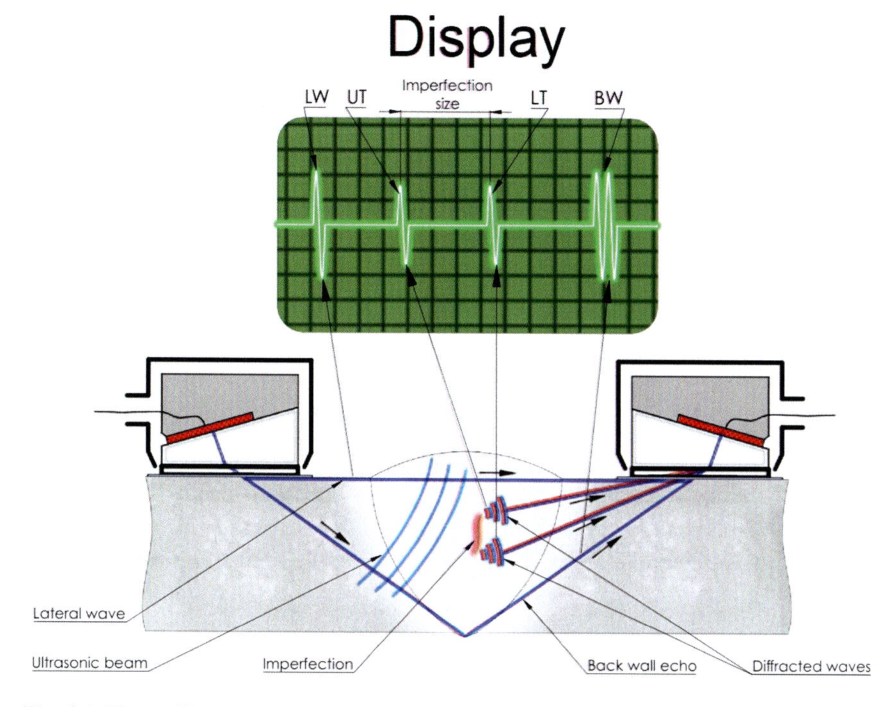

Fig. 4.6 Time-offlight diffraction technique (UT-TOFD)

In TOFD edges of imperfection are the secondary waves source. In addition, diffracted waves generated on the imperfection edges have a significant divergence angle. This defines advantages of TOFD, such as:

- High probability of defect detection in case of low requirements to precision and speed of scanning. Because of this TOFD is widely used for express testing of circular and longitudinal welds in pipelines and vessels.
- High sensitivity to small-sized imperfections.
- High precision of detection of imperfection size (as a rule, ±1 mm increasing to ±0.3 mm during re-testing).
- Imperfections can be detected regardless of their orientation.

TOFD is usually combined with UT-PE or with PAUT to test weld root and upper part of the weld.

4.3.4 Phased Array Ultrasonic Technique (PAUT)

PAUT—is the most dynamically developing ultrasonic technique with the following features:

(1) The phased array probe includes several piezoelectric elements (16–256), located on the same base plate. Each piezoelectric element is activated from the separate generator and can work independently from the others. In a common phased array probe piezoelectric plates are rectangular and are arranged in a line (see Fig. 7a). To apply current to piezoelectric plates they are covered with segmented metallic coating. In this way electric separation of elements and their independent work are ensured. Phased array probe with separate piezoelectric elements arranged as a matrix is used for better transversal beam control (see Fig. 7b). The smaller the piezoelectric element—the easier is the beam control.

(2) Physical phenomenon of wave interference is used—when waves of the same length are overlaying and as a result oscillation are generated with distribution of amplitudes equal to the sum of oscillation amplitudes in the particular location. Points where maximum values of waves are added (maximal compression of longitudinal US wave) and where minimum values are added (minimal compression of longitudinal US wave) oscillations with maximal amplitude are generated A_{max} (see Fig. 4.8).

(3) Processor-controlled input of US-pulses is used which allows to:

    (a) **focus** the ultrasonic beam at the required point by changing the phase of input of pulses from a group of piezoelectric elements in a way when each pulse reaches the control point simultaneously. *For example, to focus US beam from three piezoelectric elements according to scheme on Fig. 4.9 pulses from elements A and C must be synchronous.* Pulse from the element

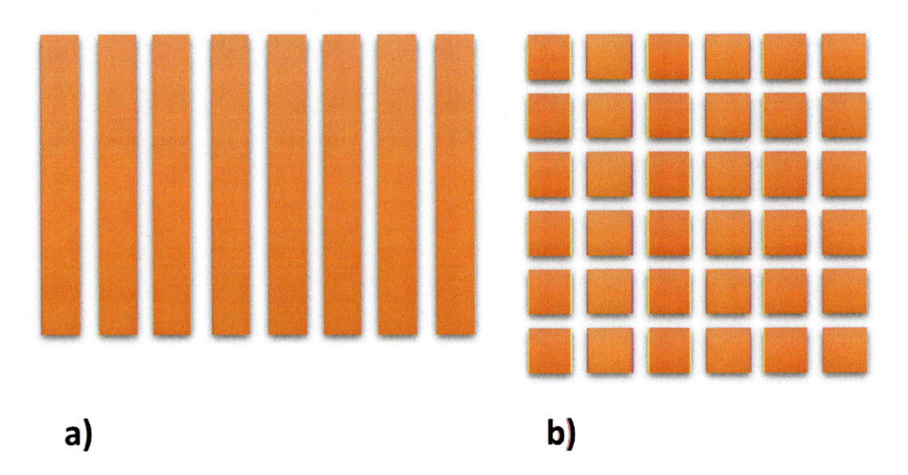

Fig. 4.7 Arrangement of piezoelectric elements in the phased array probe: **a** linear, **b** matrix

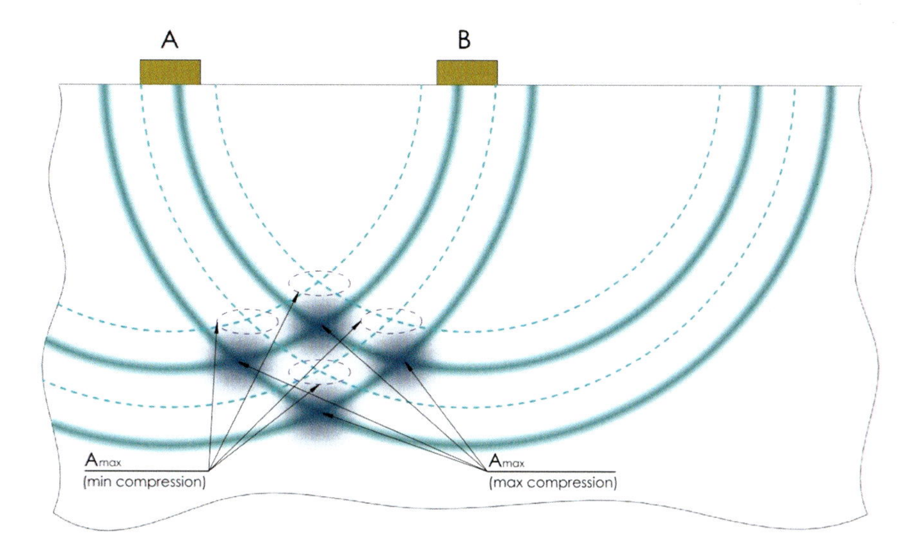

Fig. 4.8 Two-point source interference pattern

B is generated with time delay Δt_B, during which pulses from A and C travel the distance equal to $v^*\Delta t_B$ (v—sound velocity in the metal). Changing delays of input of pulses, the US-beam can be focused at any point within the control zone. In the focal point wave interference takes place with multiple amplitude increase (proportional to the number of piezoelectric elements). Accordingly, the defect echo increases. This allows to significantly improve sensitivity and resolution when imperfections are detected close to the focal point.

(b) **Adjust** the angle of probe by changing the time delays (phase) of pulses from a group of piezoelectric elements in a way when each pulse reaches the plane perpendicular to the angle of US-beam input simultaneously. In this plane wave interference takes place from different sources and a wavefront with maximal amplitude is generated. For example, for US-beam from three piezoelectric elements under the angle α (see Fig. 4.10) the pulse from A element is generated first. Pulse from B element is generated with a delay Δt_B during which the pulse from element A will travel the distance $v^*\Delta t_B$ (v—sound velocity in the metal). The pulse from element C is generated with a delay Δt_C from the pulse from B element, etc. If distances between elements are equal, delays will be equal as well: $\Delta t_B = \Delta t_C$. Any input angle of US-beam can be obtained by changing the time delays between pulses.

Input angle can be also changed by using plexiglass refracting prisms (see fig. 4.11), same as those for widely used UT-PE.

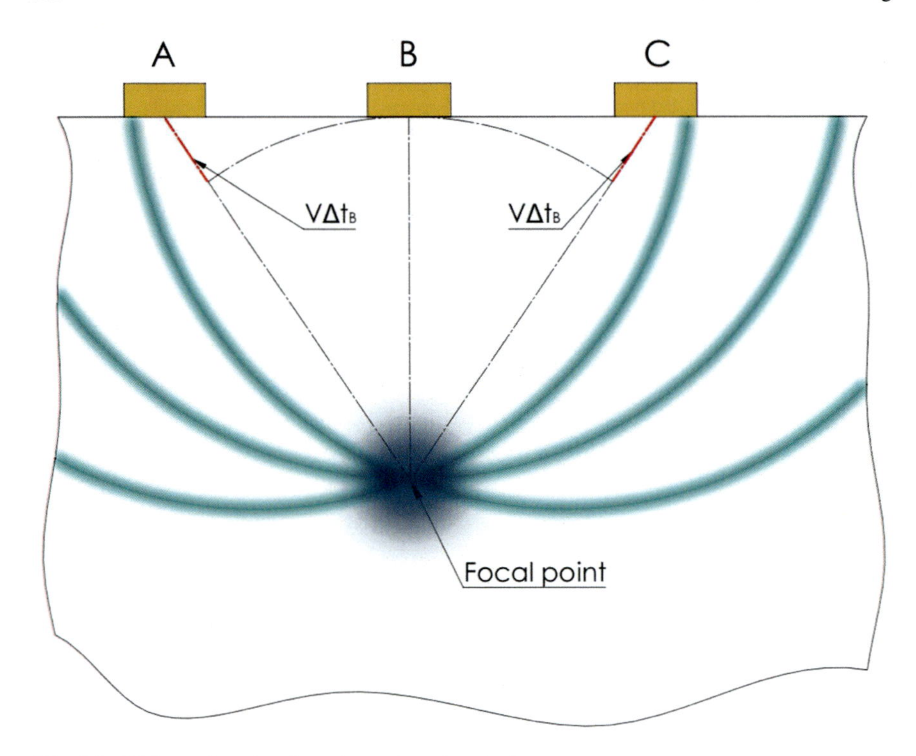

Fig. 4.9 US-beam focusing from three piezoelectric elements

(c) **Scan** the cross-section by commutation of separate piezoelectric elements or their groups in a way when the US-beam moves along the required trajectory without manipulating the probe (Fig. 4.11). Scanning significantly increases the test volume and testing productivity.

(4) Computer processing of echo is used. This allows us to obtain on the screen the so-called S-scan display **or** C-scan display (see below) as a 2D image of cross-section, to obtain 2D pattern of defect echo and to define its dimensions (see Fig. 4.11). This increases precision and eases understanding of the results.

Pattern of defect echo aside from S-scan display is also depicted as the amplitude of reflected US-pulse (A-scan display). This allows us to set the unified evaluation level for all UT techniques.

PAUT is very effective for testing simple welded structures (plates, pipes, vessels) made of low-carbon low-alloyed steels with thickness exceeding 6 mm.

PAUT is the easiest UT technique to be automated.

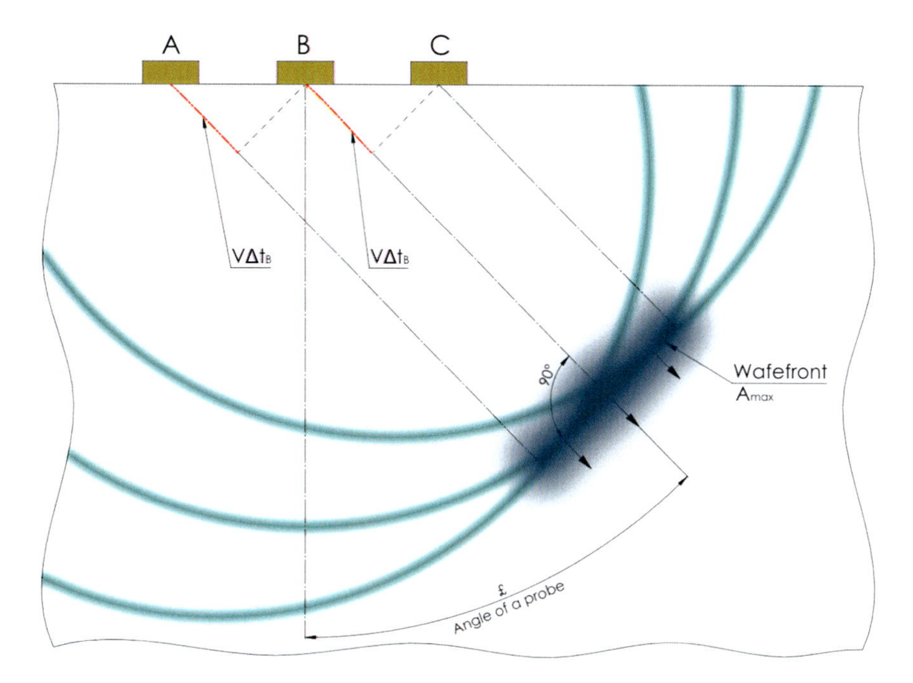

Fig. 4.10 Input of US-beam from three sources under the required angle

4.3.5 UT Technique

General stages of NDT procedures are described in Sect. 4.10.

Detailed procedures for UT are given in the following international standards:

- ISO 17640 [7]—Ultrasonic pulsed echo technique.
- ISO 10863 [8]—Time-offlight diffraction technique.
- ISO 13588 [9]—Phased array ultrasonic technique.

Main differences in UT techniques are as follows.

In UT-PE US probe is being chosen taking into consideration three parameters:

- frequency—according to the required acceptance level—2 to 5 MHz,
- angle of acoustic axis correspondent to the required angle of input of US-beam into the metal,
- size of generator—according to the distance of travelling of US in the product. The smaller the element, the lesser the length and width of the near field, the greater the beam spread in the far field for the pre-set probe frequency.

Temperature of surface of the test object should be within the range 0–50 °C.

Before welding the base metal is tested for tearing. Edges are tested within the zone with the width equal to half of total width of the weld and the HAZ.

Time based range is adjusted according to volume of control.

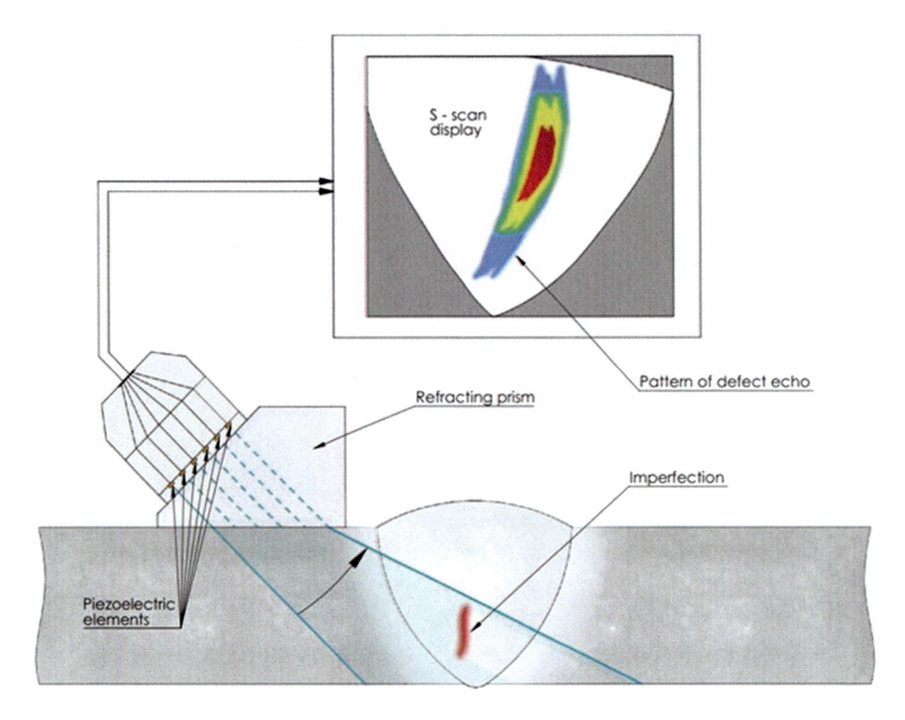

Fig. 4.11 Scanning scheme and S-scan display

Reference sensitivity level is adjusted using reference blocks with artificial imperfections. The rate of echo to noise should be at least 12 db.

Step of scanning (distance between parallel trajectories of probe movement) should be kept within the 1–3 mm range depending on product thickness. If TOFD is used the step is reduced to 0.5–2 mm. Moving the probe echoes with maximal amplitude are detected and their levels in comparison to evaluation level are noted.

If TOFD is used, probes are located symmetrically from the weld axis.

In PAUT one or several modes are to be chosen:

- B-scan—linear depiction of informative signals on the plane parallel to the US pulse direction. Is commonly used to detect depth and length of imperfections.
- C-scan—2D image obtained with scanning under constant level of US-beams input.
- S-scan—2D image obtained by sector scanning with adjustment of input angle from minimal to maximal value.

For the chosen mode focusing is performed. Focusing on beam receiving can change. This is called Dynamic Depth-Focusing (DDF). DDF increases the range of depth values on which good imperfection image can be obtained.

Scanning speed is chosen so that imperfection image was of satisfactory quality. Missing lines in the pattern indicate that the scanning speed was too high. Scanning speed is adjusted taking into consideration image resolution, pulse frequency, frequency of data collecting and volume of control. When circular welds are scanned an overlap of minimum 20 mm should be ensured between the location of the final pattern end and the beginning of the first one.

4.3.6 Advantages and Limitations of UT

Advantages of UT are:

(1) High possibility of detection of flat imperfections (cracks, lack of fusion). Method allows to detect flat imperfections more than 10^{-6} mm thick.
(2) Possibility of estimation of imperfection's depth and size which allows to apply engineering critical assessment techniques (Sect. 3.3).
(3) Method is safe for health and eco-friendly.
(4) Method allows to perform real-time control without additional data processing.
(5) Data about testing pre-sets and results are presented electronically and can be easily used as a part of Protocol of testing or Final report (Sect. 4.11).
(6) Method allows to test objects with big thickness as far as US waves can penetrate metals up to 8–10 m deep.
(7) Method is applicable for a variety of materials, metals, and non-metals alike.
(8) Low cost.
(9) High productivity.
(10) High equipment mobility.

Limitations of UT are:

(1) Sensitivity to metal structure. It is hard to test metals with coarse-grained structure, such as cast iron, because of great dissipation and attenuation of US waves. Testing of welds on austenitic steels and titanium alloys is hard as well. It is hard to test welds of dissimilar steel due to big differences between weld metal and base metal. It is almost impossible to test composite materials.
(2) Accessibility of surface for testing is needed to place the US probe.
(3) Surface should be cleaned before testing and coupling medium should be applied.
(4) "Dead zone" makes this method impossible to be used on small thicknesses. UT is effective when test object thickness exceeds 8 mm.
(5) Lower sensitivity (in comparison to the radiographic method) to volumetric imperfections (cavities and solid inclusions). UT is effective for detection of imperfections bigger than 0.5 mm long.
(6) Subjective factor—results of testing significantly depend on operator's competency and scrupulosity.

4.4 Acoustic Emission Method

4.4.1 Method Fundamentals

In NDT Acoustic emission (AE) method is used to solve three main tasks:

(a) detection of presence of developing imperfections,
(b) detection of location of developing imperfections,
(c) failure stage evaluation.

AE method is based on two phenomena:

(1) acoustic emission—allows to detect a developing defect and determine its coordinates,
(2) direct piezoelectric effect—allows to convert the elastic vibrations in the metal caused by the development of the defect into the electrical signal of the AE-sensor for subsequent amplification and analysis.

Acoustic emission—is the occurrence and propagation of elastic vibrations (acoustic waves) in the sound and ultrasonic frequency ranges due to the fast processes of energy release from localized sources (AE sources) inside or on the surface of the material.

Three types of AE are distinguished depending on the energy release mechanism:

- Material acoustic emission—acoustic emission caused by dynamic local restructuring of the material structure (see below).
- Leakage acoustic emission—acoustic emission caused by hydrodynamic and (or) aerodynamic phenomena during the flow of liquid or gas via the through hole of the test object. Acoustic waves of sound and ultrasonic frequency arise as a result of interruption of the flow and vortex formation at the edges of the through hole.
- Friction acoustic emission—acoustic emission caused by friction of the surfaces of solids.

NDT of welded structures tasks are mainly solved with material acoustic emission. Material acoustic emission sources are (see Fig. 4.12):

(a) Moving dislocations (developing plastic deformation).

    Acoustic waves arise because of an abrupt shift in the metal bond during dislocation movement.

(b) Developing crack-type imperfections of the welded joints (cracks, lack of penetration, lack of fusion, inter-dendrite shrinkage, spikes) as well as undercuts, solid inclusions, linear porosity, longitudinal cavities, and wormholes.

    Acoustic waves occur at the top of the crack as a result of:

    - developing plastic deformation,
    - ruptures of metal bonds between metal ions and energy release.

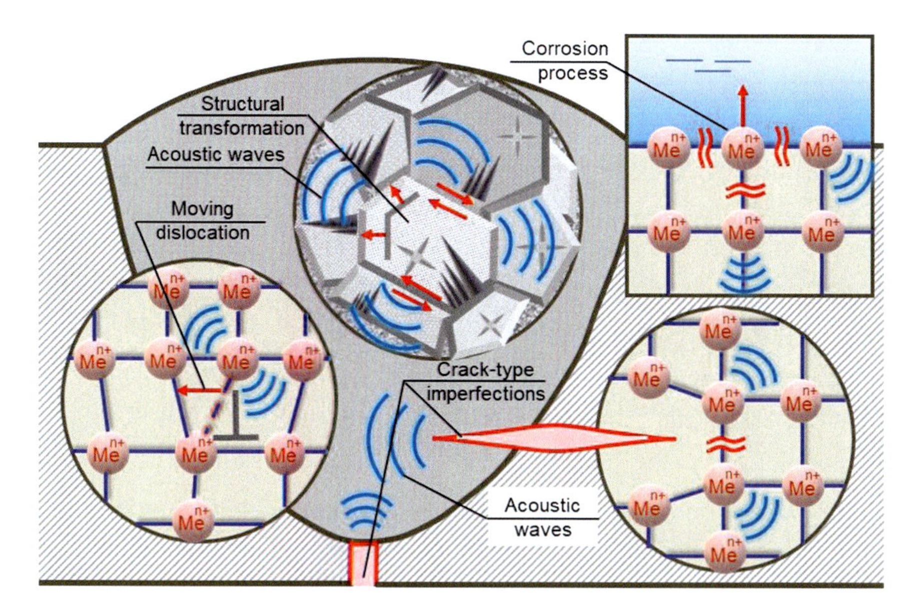

Fig. 4.12 Material acoustic emission mechanism

(c) Corrosion processes—general and local electrochemical corrosion, pitting corrosion, corrosion cracking, intergranular corrosion.
    Acoustic waves result from:

- ruptures of metal bonds upon the release of an ion from a metal into a solution during electrochemical corrosion,
- development of corrosion cracks (see Sect. 4.4.1b).

(d) Structural transformations.

Acoustic waves arise because of n abrupt increase in the crystallite volume upon a phase change (for example, during a martensitic transformation).

A single movement of an acoustic emission source (Acoustic emission event) generates AE-signal—an acoustic wave of a certain amplitude, length, and duration.

The direct piezoelectric effect is the occurrence of electric charges of different signs on the surface of the piezoelectric plate during its deformation. During tension and compression, the sign of surface potentials reverses.

During testing the AE-signal is run through the acoustic emission sensor.

For the acoustic wave to pass through the contact surface of the AE-sensor with the test object, the following requirements should be met:

- the required surface roughness (for example, achieved by grinding), since microroughnesses scatter the acoustic wave,
- the presence of a coupling medium since the liquid layer reduces the reflection coefficient at the exit point of the probe index of the ultrasonic wave.

The main element of the AE-sensor (see Fig. 4.13) is a piezoelectric plate. Mechanical vibrations of the AE-signal are transmitted to the piezoelectric plate (force P in Fig. 4.13). Compression stress arises in it with a frequency inversely proportional to the length of the acoustic wave. As a result, the potentials φ_1 and φ_2 arise on the surfaces of the piezoelectric plate. The potential difference $E = \varphi_1 - \varphi_2$ forms an electrical signal.

The electrical signal is transmitted to the AE-system for subsequent processing and visualization.

The main parameters of acoustic emission are:

- number of acoustic emission events or event count (N_Σ)—the number of registered AE-signals during the observation time,
- Acoustic emission activity (Σ)—the number of registered AE-signals per unit time,

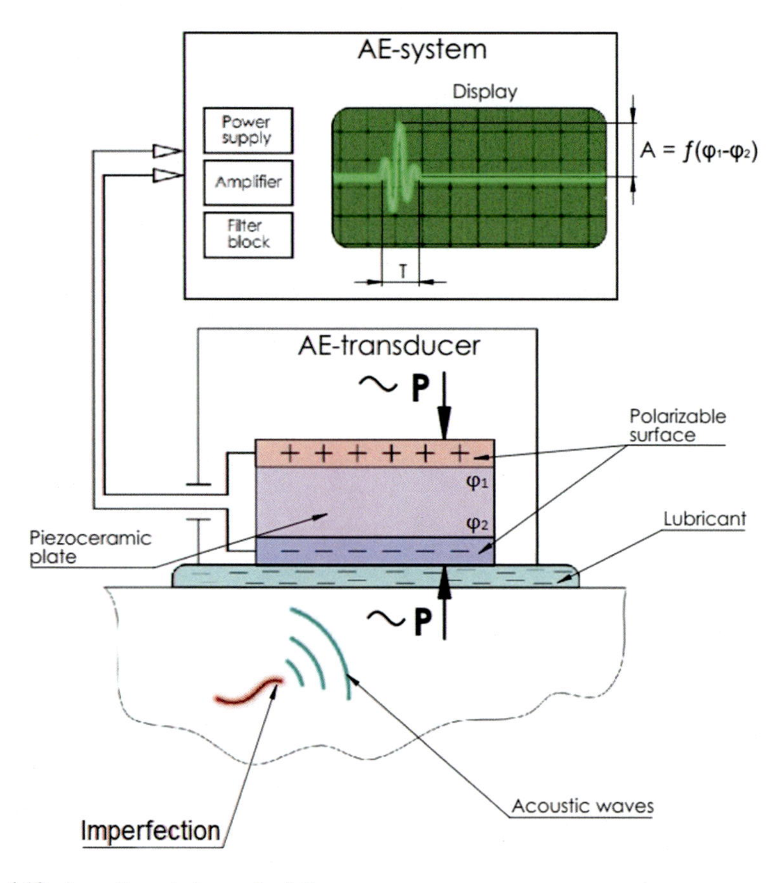

Fig. 4.13 Acoustic emission method diagram

- Acoustic emission energy (E)—the energy of an acoustic emission event. The energy of acoustic emission is characterized by the amplitude (A) and duration (T) of the electrical signal in the AE-system (Fig. 4.13).

4.4.2 Area of Application of AE

The main condition for applying the AE method is the development of defects. An imperfection can only be detected when it is an AE-source.

In this regard, the AE method can be applied:

(1) During testing of welded structures by loading, for example, hydraulic testing of vessels operating under pressure. AE-sensors are installed before loading. Defects are stress concentrators, therefore, discontinuity displacements in the defect zone occur at stresses much lower than the yield strength.

It is also possible to initiate the movement of dislocations in the flaw zone in order to identify it when monitoring a structure in service, for example, a pipeline, by slightly increasing the workload (up to 10%).

When initiating the development of a defect, it is necessary to take into account the Kaiser effect—the absence of acoustic emission in the material until the level of the previous loading is exceeded.

(2) During the operation of welded structures using stationary AE-systems. In this case, the development of defects occurs due to operational factors (load, vibration, exposure to an aggressive environment, temperature). Monitoring should be carried out continuously in real time.

The coordinates of developing defects are determined by the delay time (Δt) of the AE arrival of the acoustic emission signal from the defect to several AE-sensors. This time is determined by the AE-system.

To determine the coordinates of a developing defect in a rod or in an extended pipeline (linear coordinate system), two AE-sensors are used—No. 1 and No. 2 (see Fig. 4.14). Point 0, located in the middle (L/2) between AE-sensors, is usually used as the origin. The X coordinate of the developing defect is determined from the system of equations:

$$t1 - t2 = \Delta t \tag{4.3}$$

$$vt1 + vt2 = L \tag{4.4}$$

$$X + vt2 = L/2 \tag{4.5}$$

v *sound velocity in metal*

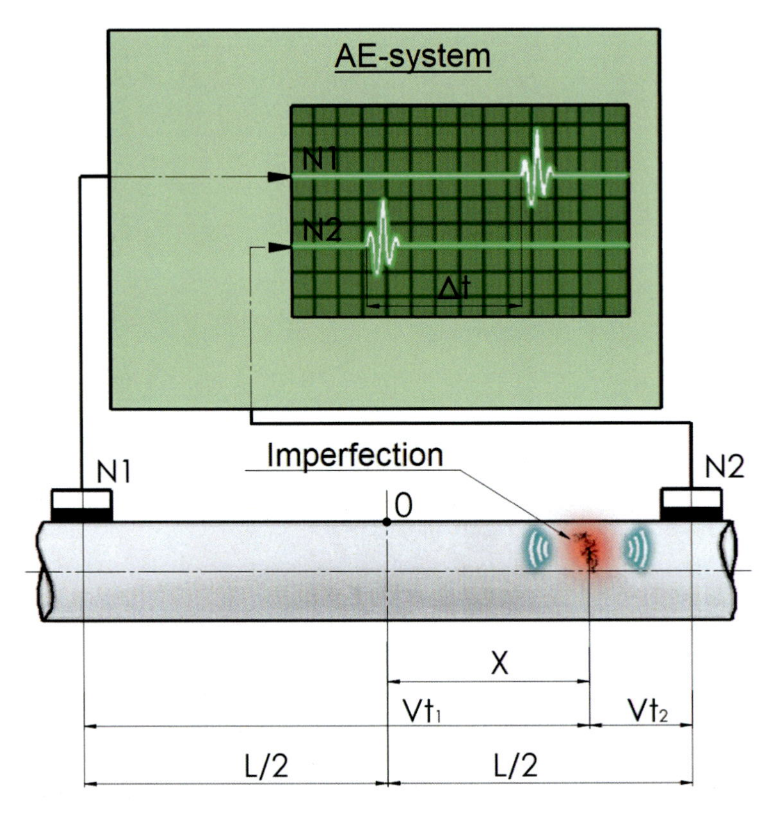

Fig. 4.14 Scheme of developing defect coordinate detection in a rod or in an extended pipeline using AE method

t_1 and t_2 *time of the arrival of acoustic wave to AE-sensors no. 1 and no. 2 respectively (unknown quantities).*

Therefore

$$X = v\,\Delta t/2 \tag{4.6}$$

To determine the coordinates of a developing defect in pressure vessels, hulls, beam elements (a flat Cartesian coordinate system), multichannel AE-systems are used. In this case, the controlled surface is divided into triangular elements. AE-sensors are installed at the vertices of the triangles.

The determination of the stages of destruction requires testing with modeling of the conditions of destruction and bringing the samples to failure.

In the simplest case of mechanical failure under static loading the AE-sources are (see Fig. 4.15):

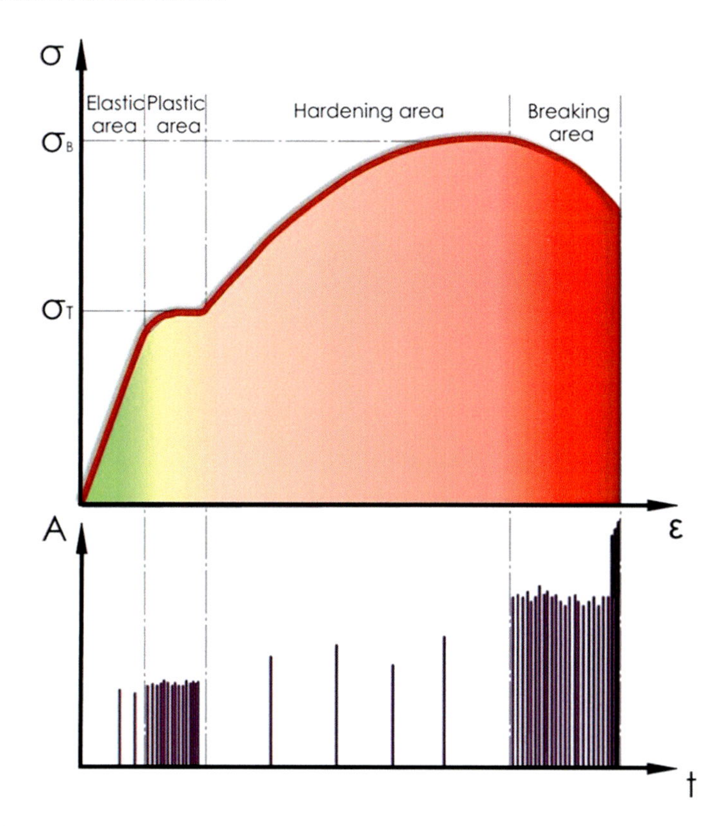

Fig. 4.15 Changes in AE-signals of different stages of imperfection under static load

- In the elastic zone—single dislocation displacements in small volumes. Acoustic emission is characterized by low acoustic emission activity and low acoustic emission energy.
- In the plastic zone—the mass dislocation movement over the entire plastic deformation zone with the dislocations reaching the surface. Acoustic emission is characterized by high acoustic emission activity and low acoustic emission energy. At the border of the plastic zone, barriers arise for dislocation movement, which leads to hardening.
- In the zone of hardening—single overcoming of barriers by dislocations. Acoustic emission is characterized by low acoustic emission activity and medium acoustic emission energy (electrical signals have a short duration and medium amplitude).
- In the fracture zone—ruptures of ionic bonds, the formation and unification of microcracks, the formation of a main crack. Acoustic emission is characterized by high acoustic emission activity and high acoustic emission energy (electrical signals have a long duration and a large amplitude).

The development of a main crack is characterized by a jump of the acoustic emission energy.

4.4.3 Acoustic Emission Technique

The general stages of NDT are described in Sect. 4.10.

The acoustic emission technique is defined by the requirements of ISO 22096 [10].

Depending on the degree of criticality of the test object, the AE method can be implemented using:

- Stationary AE-systems—acoustic emission sensors and a measuring unit are installed at the monitoring object for a long time.
- Semi-stationary AE-systems—acoustic emission sensors are installed at the monitoring object for a long time. The measuring unit is installed and connected only for the period of monitoring.
- Portable AE-systems—acoustic emission sensors and a measuring unit are installed at the monitoring object only for the duration of the monitoring.

Depending on the design features, the arrangement of AE-sensors is chosen (for example, at the vertices of triangular elements).

The surface of the structure is cleaned, and a coupling medium is applied. AE-sensors are fixed with mechanical devices (with the creation of downforce by means of a magnet, mechanical clamping, etc.) or with adhesive material. In the latter case, the adhesive material acts as a coupling medium.

An important stage is the selection of informative acoustic emission signals from the background noise of the object: noise of electronic devices (electromagnetic fields of the radio frequency range), noise from work processes (flow of liquids and gases in pipes), mechanical extraneous noise (impacts, etc.). This is achieved by applying frequency and amplitude filters.

It is effective to use the AE-method in combination with other NDT methods, for example, ultrasound. AE-method determines the presence and coordinates of the flaw, UT—its size and location.

4.4.4 Advantages and Limitations of AE-Method

Advantages of AE-method are:

(1) Obtaining data without interfering with the design of the test object.
(2) The ability to register defects at a large distance from AE-sensors, depending on the features of the welded structure and the initial amplitude of the acoustic wave. For example, the development of a crack in a main pipeline can be recorded from the distance of several hundred meters.
(3) Real-time data acquisition—without loss of time for processing the results as, for example, in the radiographic method.
(4) High sensitivity, allowing early detection of defects.

(5) The possibility to accurately determine the coordinates of an imperfection without scanning.

Limitations of AE-method are:

(1) The ability to register only developing defects. A defect cannot be detected unless it develops and there is no way to initiate its development.
(2) A significant decrease in the amplitude of acoustic waves when passing through structures of a complex spatial configuration. The scattering of acoustic waves in materials with a heterogeneous structure.
(3) The dependence of the control results from background noise.

4.5 Penetrant Testing

4.5.1 Method Fundamentals

In NDT penetrant testing (PT) is used to determine:

(a) presence of surface imperfections,
(b) imperfection coordinates,
(c) shape and size of imperfection,
(d) type of imperfection.

The PT method consists of applying a penetrating coloring liquid to the surface of the test object, penetration of the coloring liquid into the surface imperfections as a result of the capillary effect, removal of excess coloring liquid from the surface, and subsequent exposure of the coloring liquid that has penetrated into the imperfections.

The **capillary effect** is raising or lowering of liquid in tubes of small diameter (capillaries), narrow channels, porous bodies because of wetting the surface of the capillary with liquid (raising the liquid in the capillary) or non-wetting (lowering the liquid in the capillary).

Wetting characterizes the interaction of liquid molecules with surface solid molecules.

If the adhesion of liquid molecules to each other (cohesion) is weaker than the adhesion of liquid molecules to molecules of the surface of a solid (adhesion), then the liquid spreads over the surface.

If the adhesion of liquid molecules to each other (cohesion) is stronger than the adhesion of liquid molecules to molecules of the surface of a solid (adhesion), then the liquid collects on the surface in a drop.

Contact angle Θ is a quantitative characteristic of wetting. Contact angle is the angle between the surface of the body and the tangent to the surface of the liquid–gas phase (for example, to the surface of a drop).

The following main wetting options are possible (Fig. 4.16):

(1) Full wetting $\boldsymbol{\Theta = 0^\circ}$. The liquid spreads over the surface to a monolayer (for example, gasoline or oil on the surface of the water).

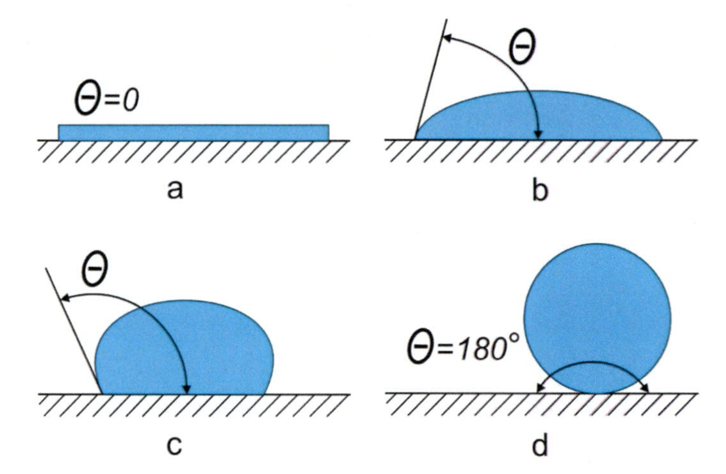

Fig. 4.16 Liquid wetting of a solid surface: **a** complete wetting; **b** strong wetting; **c** weak wetting; **d** complete non-wetting

(2) Strong wetting $\mathbf{0° < \Theta < 90°}$. A surface with strong wetting is called wettable (for example, a drop of alcohol on a glass surface).
(3) Weak wetting $\mathbf{90° < \Theta < 180°}$. A surface with weak wetting is called non-wettable (for example, a drop of water on the surface of paraffin).
(4) Complete non-wetting $\mathbf{\Theta = 180°}$. The fluid collects in a sphere (for example, a drop of mercury on a wooden surface).

If a capillary is placed in the wetting liquid, the adhesion forces lift the liquid along the capillary to a height **h** (Fig. 4.17) until it is balanced by the force of gravity acting on the liquid column in the capillary. Via this mechanism, the coloring liquid deposited on the surface of the test object penetrates the surface imperfections.

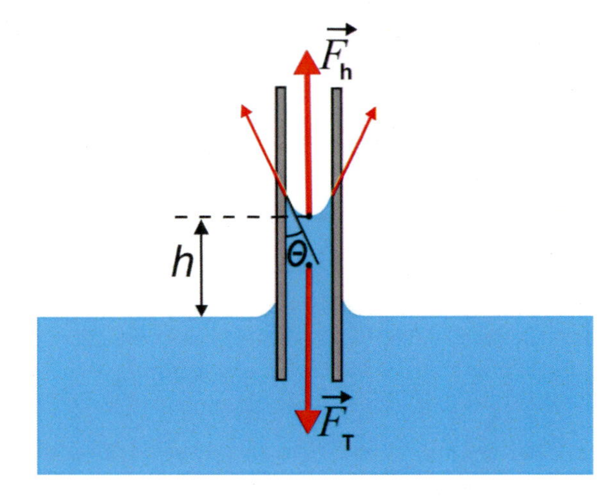

Fig. 4.17 Lifting of the wetting liquid through the capillary

4.5.2 Applicability of PT Method

The capillary method is used to control surface imperfections in castings, forgings, welded joints from any materials.

The capillary method is classified according to two criteria.

(1) By the type of penetrating coloring liquid (penetrant):

- The method of penetrating solutions—penetrant is a liquid indicator solution.
- The method of filtering suspensions—penetrant is a liquid suspension, which forms an indicator pattern of filtered particles of the dispersed phase.

(2) By the method of visualization of the indicator pattern on the background of the surface of the test object:

- luminescent method—a luminescent indicator pattern is recorded when illuminated with long-wave ultraviolet light.
- color method—a color indicator pattern is recorded under normal lighting.
- luminescent-color method—a luminescent or color indicator pattern is recorded when illuminated with long-wave ultraviolet light.
- The brightness method—the achromatic (black-and-white with various shades) indicator pattern is recorded under ordinary lighting.

4.5.3 PT Technique

The general steps of NDT are described in Sect. 4.10.

Details of PT procedures are given in the international standard ISO 3452–1 [11].

In the PT technique the main steps are as follows (see Fig. 4.18).

(a) Preparation and preliminary cleaning of the surface of the test object.

Contamination (paint, scale, rust, oil) must be removed mechanically using a brush, sandpaper, sanding, shot and sandblasting, high pressure water jet cleaning, etc. After cleaning, the test area should not be closed due to plastic deformation or clogging with abrasive materials. The final pre-cleaning operation is drying of the controlled surface.

(b) Application of penetrant.

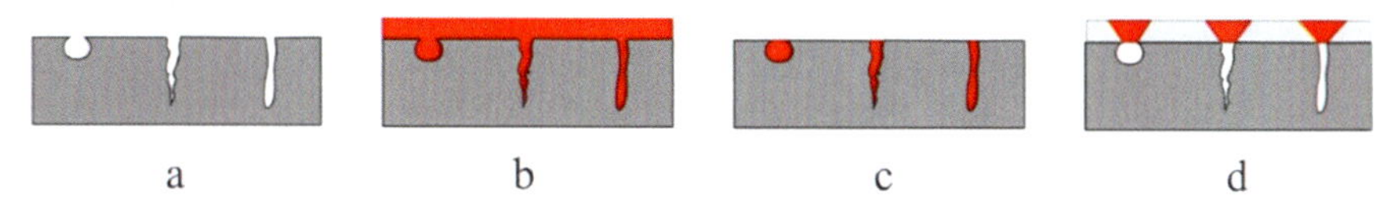

Fig. 4.18 The main stages of the capillary control technique: **a** preparation of the product, **b** application of penetrant, **c** intermediate cleaning, **d** exposure

The penetrant is applied to the test object by spraying, brushing, watering, or dipping. The penetrant exposure time is regulated by the manufacturer (from 5 to 60 min).

(c) Intermediate cleaning (removal of penetrant from the surface) is carried out by washing, dipping, or wiping with water. Next, the surface is dried. The surface temperature during drying should not exceed 50 °C so that the penetrant that has got into surface imperfections does not dry out.

(d) Exposure—pulling penetrant from surface imperfections for visualization purposes. For this purpose, a special solvent-based substance is used—the developer. The developer is applied in a uniform thin layer to the surface by spraying: with compressed air, electrostatic, "boiling" layer or in a swirl chamber. After exposure, the excess developer is removed.

Inspection is carried out to identify imperfections. When using color penetrants, the illumination should be at least 500 lx. When using fluorescent penetrants, inspection is carried out in the dark with background ultraviolet light.

The final cleaning of the product after inspection is carried out if the penetrant and developer used can affect the further production or service processes.

4.5.4 Advantages and Limitations

The advantages of PT are:

(1) High performance.
(2) Applicability for any type of materials.
(3) Simplicity and cheapness.
(4) High information content, the ability to tentatively determine the shape and size of imperfections.
(5) Mobility, the possibility of use both in laboratories and in production shops, and on industrial sites during the structure's installation.

The limitations of the PT are:

(1) Ability to control only surface imperfections.
(2) Dependence of imperfections detection on the quality of preparation of the test object surface.
(3) Health hazard and fire hazard of flammable and/or volatile substances used.
(4) Dependence of the results of control on the competence of the controller (subjective factor).

4.6 Magnetic Particle Testing

4.6.1 Method Fundamentals

In NDT Magnetic particle testing (MT) is used to identify:

(a) presence of surface and subsurface discontinuity-type imperfections of ferromagnetic materials—cracks, cavities, solid inclusions, non-fusion (respectively 1, 2, 3 and 4 groups of imperfections—Sect. 3.1),
(b) imperfection coordinates,
(c) tentatively, the shape and dimensions of the imperfection.

The method is based on magnetizing the ferromagnetic materials with an external magnetic field and indicating the magnetic fluxes of dispersion that occur on the surface of the product in the imperfection zone.

(1) Magnetization of ferromagnetic materials by an external magnetic field—the formation of a total magnetic field due to the directional orientation of the elementary electromagnets (domains) that make up the ferromagnetic materials—see Sect. 4.8.1, Fig. 4.25.
(2) Magnetic scattering fluxes are formed due to different magnetic permeabilities of steel (relative permeability of carbon steels is ~100) and imperfection material (relative magnetic permeability of gases, of flux and tungsten inclusions is ~1). In the absence of an imperfection, the magnetic flux of the magnetic field of the magnetization closes completely through the control object. The imperfection is an obstacle for the magnetic flux of the magnetic field magnetization. The magnetic flux density at the boundaries of imperfection increases. If the boundary of the imperfection is near the surface of the test object, part of the magnetic flux extends beyond the surface—a magnetic flux of scattering is formed (Fig. 4.19). The scattering magnetic flux increases with:

    - approaching the imperfection to the surface.
    - increasing the length of the imperfection.
    - sharpening the boundaries of the imperfection (maximum scattering fluxes form over the cracks, minimal over the pores).
    - orientation of the maximum imperfection size perpendicular to the magnetic flux of the magnetization magnetic field.

(3) Indication of magnetic fluxes of dispersion is carried out due to the accumulation of magnetic particles (powder of ferromagnetic material) because of magnetization by a magnetic field of dispersion. During testing, magnetic powder is poured as an even thin layer in the magnetization zone. In the presence of imperfection, the scattering field magnetizes the powder. Magnetized powder forms clusters—an imprint (Fig. 4.19), the shape and size of which corresponds to the shape and size of the dispersion flux.

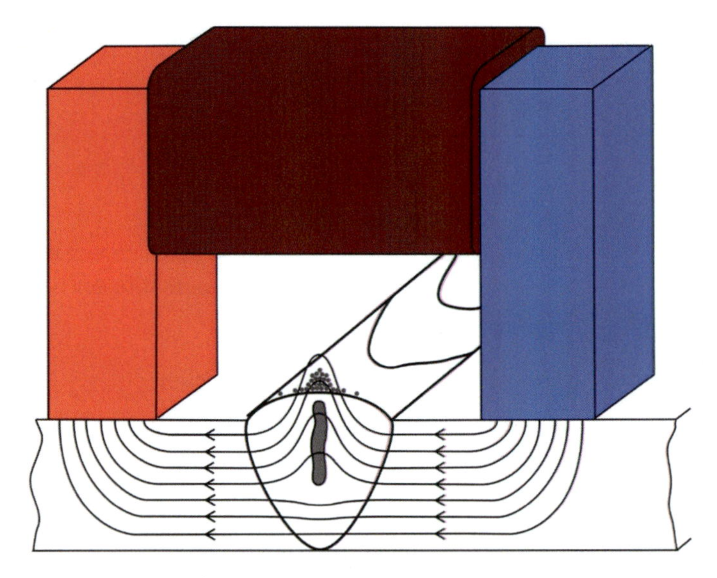

Fig. 4.19 Diagram of the formation of a magnetic flux of dispersion and accumulation of magnetic powder over a subsurface imperfection (imprint)

4.6.2 Applicability of Magnetic Particle Testing

MT is used to control surface and subsurface (with a depth of up to 2 mm) imperfections in castings, forgings, and welded joints made of ferromagnetic materials.

4.6.3 MT Technique

The general steps of NDT are described in Sect. 4.10.

Details of magnetic particle inspection procedures are given in the international standard ISO 17638 [12].

The main features of the magnetic particle inspection technique are as follows.

The preparation of the control surface involves primarily the removal of moisture. It is possible to clean the surface with sandpaper. The surface can be coated with a thin (up to 50 micron) layer of non-magnetic paint. With a larger coating thickness, the sensitivity of the method decreases.

Magnetization is carried out in one of three ways:

(a) Yoke electromagnet (Fig. 4.20) or permanent magnets—the control zone becomes part of the patch magnetic circuit. The method allows to create a magnetizing field with adjustable induction to large values.

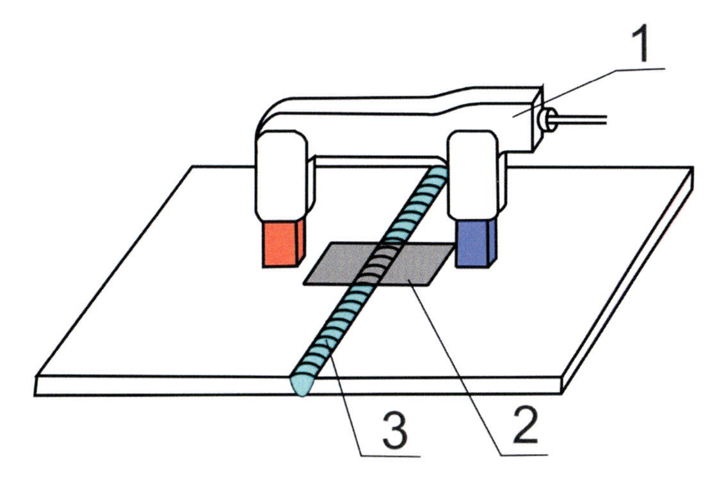

Fig. 4.20 The scheme of magnetization of a welded joint with a yoke electromagnet (ISO 17638): 1—yoke electromagnet, 2—magnetization zone, 3—weld

(b) Current generators with laid on electrical contacts (Fig. 4.21)—alternating current from the generator is passed through the test zone. Alternating current creates a magnetizing electromagnetic field. The current supply is carried out by overhead electrical contacts.

(c) Current conductors wound around the product (Fig. 4.22)—the control zone becomes part of the magnetic core. The method allows to simply and efficiently create a magnetizing field with adjustable induction to large values. However, it is applicable only for cylindrical test objects (mainly pipelines).

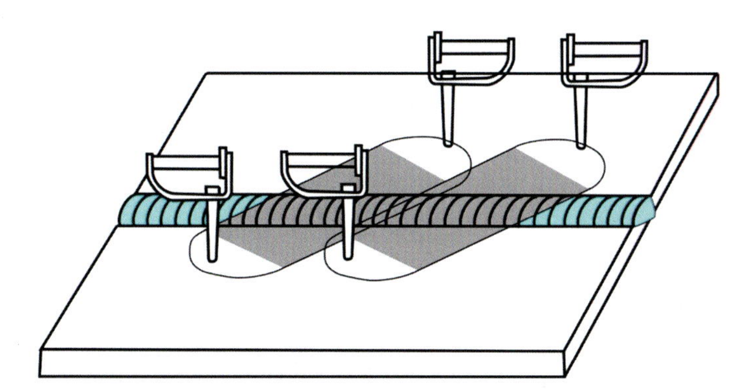

Fig. 4.21 The circuit of magnetization of a welded joint by current generators with laid on electrical contacts—ISO 17638 (the diagram shows two positions of laid on electrical contacts, providing overlapping test zones)

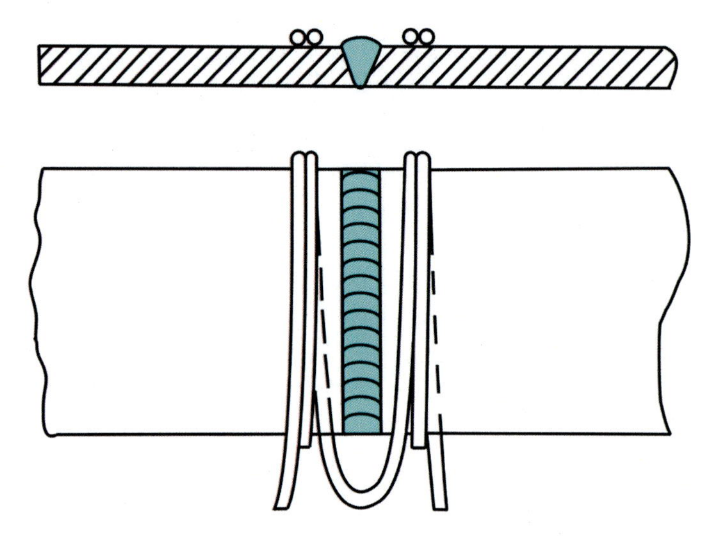

Fig. 4.22 Magnetization scheme with current conductors wound on the product (ISO 17638)

Magnetization verification is carried out by one of the following methods:

- identification of an artificial or natural imperfection in the control sample.
- measurement of the tangential component of the magnetic field by the Hall sensor.
- calculation of the average value of the current providing the tangential component of the magnetic field strength from 2 to 6 kA/m.

The magnetization of the welded joint is carried out:

- according to the scheme, providing magnetization of the weld with the heat-affected zone plus 50 mm.
- in two mutually perpendicular directions.

A change in the direction of magnetization is carried out because the detectability of imperfection increases significantly when the magnetic flux of the magnetization field is oriented perpendicular to the maximum size of the imperfection.

The application of a magnetic indicator is done via one of the options:

- sprinkling powder
- spray
- watering with an emulsion.

Registration of the obtained imprint of the scattering field from the imperfection is carried out using the following:

- photo or video recording,
- written description,
- drawing,
- magnetic tape,

- fixation of the magnetized imprint (transparent adhesive tape, varnish, peelable paint, epoxy).

Demagnetization of the test object is carried out only if necessary (since the residual magnetization after control is low) to a level of residual magnetization of no more than 0.4 kA/m.

4.6.4 MT Advantages and Disadvantages

The advantages of MT are:

(1) The simplicity and cheapness of the method.
(2) High information content, the ability to tentatively determine the shape and size of imperfections.
(3) Mobility, possibility of use in laboratories, in production shops, and on industrial sites during the installation of structures.

The limitations of MT are:

(1) Dependence of imperfection detection on depth, orientation relative to the magnetic flux of the magnetic field of the direction, size, and shape of the imperfections. It does not reveal deep-seated internal defects.
(2) Applicability to ferromagnetic materials only. Alloying of steels, as a rule, reduces the specific magnetic permeability (exceptions are, for example, ferritic and martensitic stainless annealed steels). This requires an increase in the magnetization field strength. The method is not applicable for the control of aluminum, titanium, copper, and their alloys.
(3) The need for surface preparation.
(4) The dependence of the results of control on the competence of the controller (subjective factor).

4.7 Eddy Current Method

4.7.1 Method Fundamentals

In non-destructive testing, the Eddy Current Method (ET) is used to determine:

(a) the presence of surface and subsurface defects that are discontinuities—cracks, cavities, solid inclusions, non-fusion (respectively groups 1, 2, 3 and 4 of imperfections—Sect. 3.1),
(b) imperfection coordinates
(c) metal structure changes.

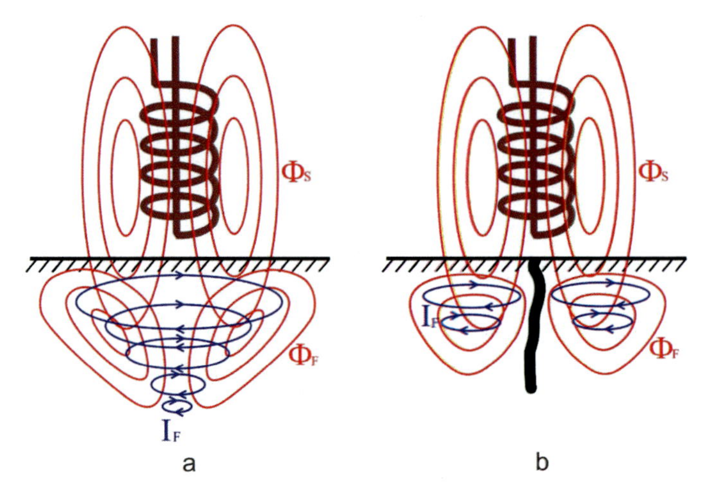

Fig. 4.23 Schematic of eddy currents: **a** in metal without imperfections, **b** in cracked metal (Φ_S—magnetic fluxes of the exciting coil of the eddy current converter, I_F—eddy currents, Φ_F—magnetic fluxes of eddy currents)

The ET method is based on the physical effect of eddy currents appearance (or Foucault currents)—induced 3D electric currents in a ferromass when the magnetic flux of an external magnetic field acting on it changes with time.

As a result of electromagnetic induction, the ferromass electrons begin to move around the lines of force of an external alternating magnetic field and create closed cyclic (eddy) currents. The electric circuit of the eddy current creates its own magnetic field. The magnetic flux of this field is directed parallel to the magnetic flux of the main external magnetic field, but in the opposite direction. Thus, the magnetic field of the eddy current prevents a change in the main external magnetic field.

When ET is performed in the absence of imperfections, the following occurs (Fig. 23a):

(1) The exciting coil of the eddy current converter generates an alternating magnetic field. Under the action of the field in the controlled product, an alternating magnetic flux Φ_S occurs.
(2) The variable magnetic flux Φ_S induces eddy currents I_F in the ferromass (tested product). Eddy currents flow in the tested product along concentric annuli circles.
(3) Eddy currents create their own magnetic field. Moreover, the magnetic flux Φ_F of the magnetic field of the eddy currents is directed opposite to the magnetic flux Φ_S of the exciting winding of the eddy current converter.
(4) The magnetic fluxes of the exciting coil of the eddy current converter Φ_S and eddy currents Φ_F are subtracted. In this case, the inductive resistance of the exciting winding of the eddy current converter changes.

In the event of an imperfection, the eddy current configuration I_F changes. So, in the presence of a crack (Fig. 23b), there are several zones of eddy current formation.

This leads to a change in the magnetic flux Φ_F of the magnetic field of the eddy currents.

As a result of changes that occur in the tested metal, including the presence of imperfections, the interconnected electromagnetic characteristics of the system "windings of the eddy current converter—ferromass of the test object" change:

- inductive resistance and impedance (complex resistance with active, inductive, and capacitive components) of the exciting winding of the eddy current converter.
- EMF and voltage of the exciting winding of the eddy current converter,
- inductance of the exciting winding of the eddy current converter,
- magnetic field strength of the test object,
- electrical conductivity of the test object,
- relative magnetic permeability of the test object,
- other electromagnetic characteristics.

In addition, each electromagnetic characteristic reacts differently to changes in the characteristics of the test object:

- presence of imperfections,
- type of imperfection,
- material,
- structure,
- design features (shape, size, presence of flanges, ribs, etc.),
- types of welded joints.

In ET, two main types of eddy current converters are used:

(1) parametric—structurally consists of one winding, which acts as both exciting and measuring one,
(2) transformer—structurally consists of two windings: exciting and measuring.

In addition, eddy current converters are divided into:

- Overhead—windings are applied to the test object. This is the main type of eddy current converters for monitoring of welded structures.
- Pass-through: external—the test object moves inside the windings (for example, control of small diameter pipes, wires, etc.) or internal—the windings move inside the test object (for example, during in-line diagnostics).
- Combined and other types.

The parameters of the ET are:

- The amplitude of the electric signal induced in the measuring winding (for transformer eddy current converters), or the change in voltage in the magnetizing winding as a result of a change in inductive and active resistance (for parametric eddy current converters).
- The phase of the electrical signal.
- Impedance diagram of eddy current probe (Fig. 4.24)—a graphical representation of the complex resistance changes in Cartesian coordinates with the "active

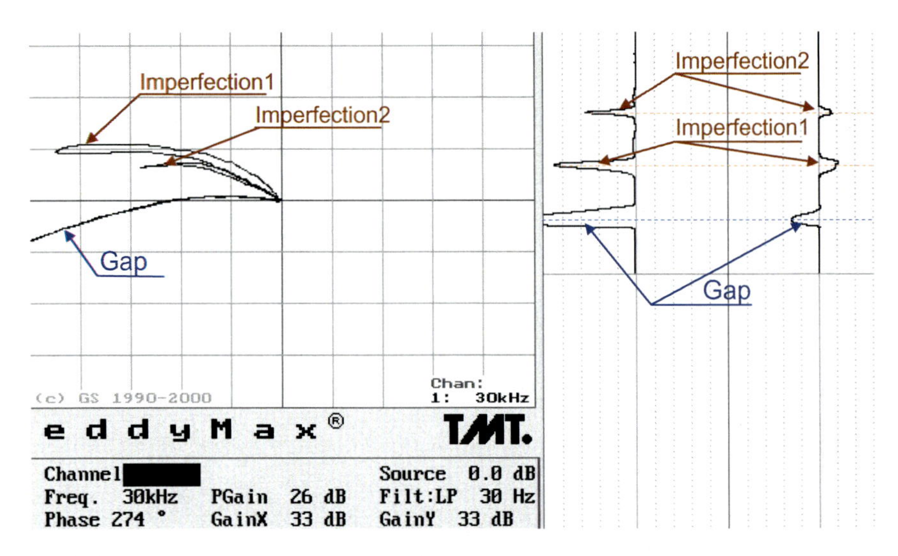

Fig. 4.24 An example of the interface of an ET defect detector in the presence of two imperfections (and the gap between the transducer and the surface): left) impedance diagram of eddy current probe, right) sweep of the active resistance and inductive resistance

resistance—inductive resistance" axes obtained as a result of changes in the frequency, electrical conductivity, relative magnetic permeability, characteristics of the control object or their combination.

- Other electromagnetic characteristics.

4.7.2 ET Area of Application

According to ISO 17643 [13], the ET method is used to control products from the following materials:

- steel of all types,
- aluminum and aluminum alloys,
- titanium and titanium alloys,
- copper and nickel-based alloys.

The ET method is used in the following industries:

- aviation (airframe designs, including skin, turbine engine blades, wheel disks, etc.),
- railway transport (parts and units of wagons, wheelsets, axle boxes, freight trolleys, passenger cars, rails, etc.),
- oil, gas and chemical (pipelines, gas turbine blades, pressure vessels, reservoirs for storing liquids, etc.),

- power engineering (pipes of heat exchangers, steam pipelines, collectors, blades of steam turbines, etc.),
- mechanical engineering (rods, wires, sheet metal, welded joints of hull structures, etc.).

4.7.3 Eddy Current Testing Techniques

The general steps of non-destructive testing are described in Sect. 4.10.

The features of the BT method are:

- a large number of control parameters, each of which determines sensitivity of testing to various characteristics of the test object, including types of imperfections,
- a large number of designs of eddy current converters.

These features determine the wide possibilities of the ET method and the need to develop separate procedures for various combinations of characteristics of the test object listed in Sect. 4.7.1.

The general procedures for all techniques of the ET method are as follows.

Hardware setup—selection of the operating monitoring frequency (magnetization current frequency) and amplification (magnetization current value), at which a response to artificial imperfections of reference samples is provided. Typically, an operating monitoring frequency of 50 Hz to 2 MHz is used. The greater the operating monitoring frequency f, electrical conductivity б, and magnetic permeability of the material of the part, the smaller the depth at which eddy currents can be induced in the metal:

$\delta = 1/\sqrt{(\pi f б \mu_0)}$ (Sect. 4.7.1).

δ—the depth at which the magnitude of the eddy currents decreases by e times (the detection of imperfections remains almost constant at a depth of up to 3δ),

$\mu_0 = 4\pi 10^{-7}$ *H/m—relative magnetic permeability of the material*

Depending on the types of controlled defects, various samples are used as standards. Samples with a 0.5–2 mm depth notch and no more than 0.2 mm width are used often. The notch should give a signal of about 80% of the height of the screen.

The ET method does not require surface preparation. Testing is possible through non-metallic coatings up to 2 mm thick.

Scanning of the surface determined by the control volume is carried out separately for the surface of the weld, heat affected zone and base metal. When scanning, the converter is moved along a zigzag path with a step that allows to overlap adjacent magnetization zones.

The imperfection coordinates are fixed by the location of the eddy current converter with control parameters exceeding the acceptance level.

When monitoring changes in the metal structure, control parameters are compared with the range of values obtained on samples or products with the required structure.

4.7.4 Advantages and Limitations of ET

The advantages of the ET method are:

(1) High detectability of surface and subsurface defects.
(2) High performance.
(3) The simplicity and cheapness of the method.
(4) Mobility (can be used in laboratories, production shops, during the installation of structures)
(5) The possibility of contactless testing. This allows testing to be carried out during the mutual movement of the converter and the test object with high speed. For example, during ET of rail transport, rail monitoring is carried out at speeds of up to 70 km/h.
(6) Versatility. A large number of informative testing parameters makes it possible to control any changes in the characteristics of the test object.
(7) Ample opportunities for automation and the creation of multi-channel systems for testing similar products.

Limitations of the ET method are:

(1) The possibility of using only for structures made of electrically conductive materials.
(2) The inability to detect deep-seated internal imperfections. The depth of defect detection is limited by the depth of propagation of the magnetic flux of the magnetic field of the exciting coil of the eddy current converter.
(3) The inability to determine the imperfection size. The readings from a shallow defect on a surface or shallow depth are the same as the readings from a large defect at a greater depth.
(4) The need to remove the bulge to control the weld.
(5) The subjectivity of manual ET—the results depend on the competence of the controller.

4.8 Magnetic Anisotropy Method

4.8.1 Method Fundamentals

In non-destructive testing, the magnetic anisotropy (MA) method is used to determine the magnitude of deformations (or stresses in the elastic range) in carbon and low-alloyed steels.

The MA method is based on the use of a **magnetoelastic effect** (Villari effect)—the change in the magnetic characteristics of ferromagnetic materials as a result of deformation.

The magnetoelastic effect is the opposite of the magnetostriction effect (Joule effect)—the change in the linear size and volume of ferromagnetic materials when the state of magnetization changes.

Carbon low-alloyed steels are ferromagnetic. Their structure at the grain level consists of domains—groups of ions in the nodes of the crystal lattice, which together with the surrounding electron cloud are elementary electromagnets. In the initial state, the magnetic fields of the domains are oriented chaotically and are therefore compensated. The total magnetic field, which is characterized by the magnetization vector, is created by the domains under the condition of their directional orientation (amplification of the magnetic field of domains of a certain orientation). Directional domain orientation can occur:

(1) *Under the influence of an external magnetic field*. The magnetization is carried out by changing the phase volumes (shifting the domain boundaries) and changing the momentum ("rotation") of the magnetization (see Fig. 4.25). In small magnetic fields, the change in magnetization occurs mainly by shifting the boundaries. Next is the range of magnetic fields, in which the boundary displacement and rotation occur simultaneously, in even larger fields exclusively rotation processes are present. Changing the domain boundary is paused at the grain boundaries and crystal lattice defects. Further advancement of the domain boundary requires more energy to jump over the grain boundaries.

(2) *As a result of metal deformation*. Furthermore, the tensile and compressive stresses have an influence on the magnetic domains' alignment in a different way as presented schematically in Fig. 4.26. A total magnetic field of directionally oriented domains is created and the magnetic permeability changes.

Tasks of MA method in NDT are:

(a) to determine the change in magnetic permeability due to deformation,

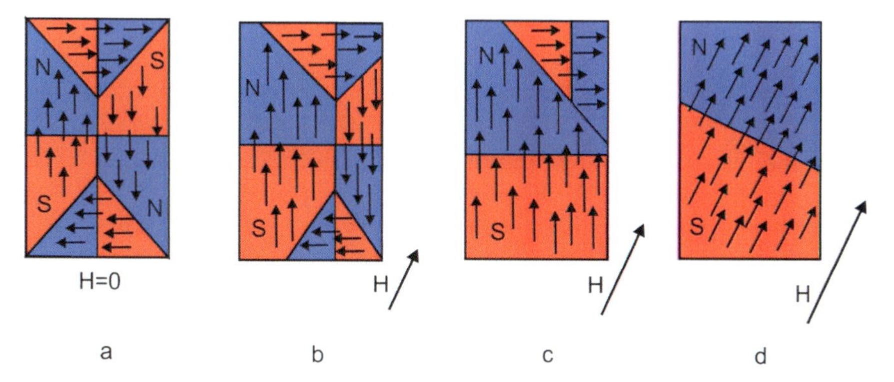

Fig. 4.25 Stages of magnetization of ferromagnetic material: **a** without external field, **b** weak field, **c** strong field, **d** saturation (single domain state)

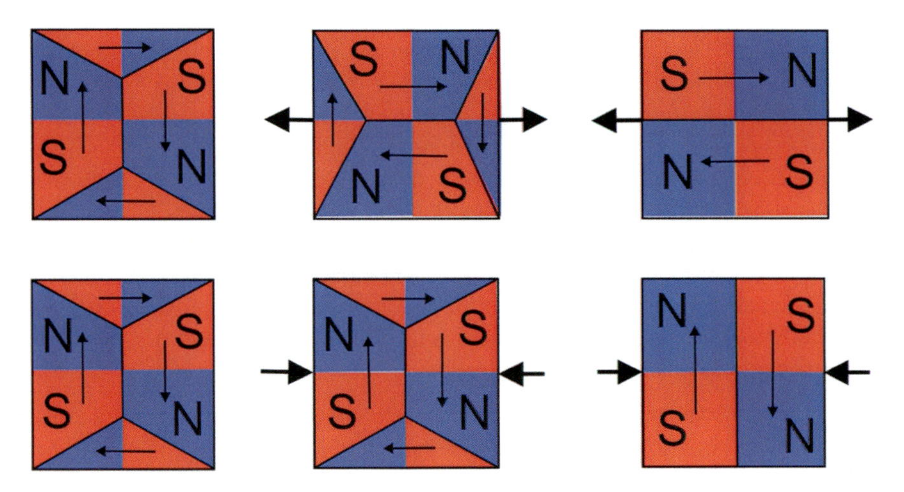

Fig. 4.26 Rearrangements of magnetic domains under the influence of tensile (above) and compressive (below) deformation

(b) to establish a quantitative relationship between the change in magnetic permeability and the magnitude of deformations (or stresses in the elastic range)—calibration of the MA test instrument.

The change in magnetic permeability is determined by:

(3) creating an alternating magnetic field in the control zone of the excitation coil,
(4) indication of changes in magnetic flux by sensing coil.

The different types of the four-pole probes are applied as primary sensors in MA stress measurement techniques. Usually the probes are composed of the two coils mounted on the two separated U-shaped (or yoke type) cores, which are positioned perpendicularly to each other. One coil (excitation coil) commonly relates to the harmonic generator output. The other coil is a sensing coil which is linked to the measuring block. The probes interact with testing surface and are arranged as a 4-pole system where sensors are in the corners of a regular tetragon (see Fig. 4.27).

When an alternating current is applied to the excitation coil W1 (poles H–H Fig. 4.27), an alternating magnetic field is created in the testing area. This results in the appearance of a magnetic flux Φ that connects the poles of the H–H excitation coil. In the measuring coil sensing coil W2 (poles B-B Fig. 4.27) an output electric signal U is generated, proportional to the difference of the components of the magnetic fluxes $\Phi1-\Phi2$.

When stresses are absent the magnetic fluxes are compensated: $\Phi1 = \Phi2$. Output electric signal $U = 0$.

Under the action of stresses, the magnetic permeability in the direction of stresses increases (in the case of tensile stresses) and decreases in the transverse direction. As a result, the magnetic flux $\Phi1$ increases, and the magnetic flux $\Phi2$ decreases. An output electrical signal $U \neq 0$ occurs. In uniaxial stress state, the output electrical

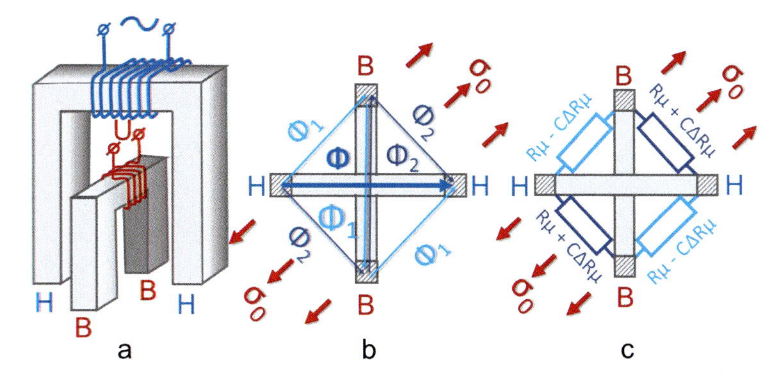

Fig. 4.27 Scheme of 4-pole probe: **a** structure, **b** magnetic fluxes, **c** magnetic bridge

signal is proportional to the voltage (in the elastic range). In the biaxial stress state, the output electrical signal is proportional to the difference of the main stresses.

The MA probe work can also be explained using a magnetic bridge (Fig. 27c). The testing zone between the probe poles arranged as a square is a magnetic bridge with zones of surface acting as its branches. Under the tensile load the magnetic penetration increases. As a result, the magnetic resistance Rμ reduces its value by CΔRμ. In the perpendicular direction, the magnetic resistance Rμ increases by CΔRμ. In this way on the bridge diagonal along the measuring coil W2 the signal U, proportional to the difference of magnetic resistances of the perpendicular branches of the bridge, is formed.

Due to high enough operational frequencies the MA methods can be considered as one of the versions of the EC method.

4.8.2 MA Method Area of Application

MA method is applied:

(1) In the process of welded structures design—to select a productive assembly and welding technology for solving the problem of minimizing residual welding stresses. In this case, the distribution of residual welding stresses in the control sections of structures made, for example, with different assembly and welding sequences, is determined.
(2) During the manufacturing of welded structures—to monitor the effectiveness of post-welding treatment to reduce residual stresses (for example, heat treatment). In this case, the distribution of residual welding stresses in the control sections of the structures before and after post-welding treatment is determined.
(3) In the process of welded structures operation—for monitoring the most important zones as those with the highest probability of failure to take actions

preventing it. In this case, the change in time of the distribution of residual welding stresses in the control sections of the structures is determined.

4.8.3 MA Technique

General procedures of NDT are given in Sect. 4.10.

MA technique features are the following.

Determining the initial magnetic anisotropy. Construction materials are almost always anisotropic in magnetic properties because of manufacturing technologies (rolling, bending, etc.). Due to the initial magnetic anisotropy, the MA test instrument will show a certain value in the absence of mechanical stresses. To consider the initial magnetic anisotropy, the initial reading of the MA test instrument on the unloaded section of the structure is carried out. When controlling stresses, the initial reading is subtracted from the measured value.

MA test instrument calibration—determination of the dependence of the readings of the instrument obtained as a result of measurements on the magnitude of the stresses.

The calibration is carried out with uniaxial tension on the tensile machine. For calibration, a flat sample made of steel of the same group as the controlled structure is used. The sample is loaded in the elastic range. Actual stresses are defined as the ratio of the load to the cross-sectional area of the sample. In the process of loading the readings of MA test instrument are determined.

According to the calibration results the calibration coefficient T is determined as the average increase of stresses per unit reading:

$$T = 1/N \, \Sigma \, (\Delta \sigma_i / \Delta A_i) \tag{4.7}$$

ΔA_i *increase of instrument reading corresponding to* $\Delta \sigma_i$ *stresses increase,*
N *number of measurements performed during loading.*

The surface of the structure is cleaned from dirt and rust in the control area. Measurements on non-metallic coatings (paint) up to 0.3 mm thick are permissible. If insulating coatings are used, they must be removed.

Processing of measurement results. In case of uniaxial stress state, the stresses are determined by the ratio:

$$\sigma_1 = T(A - A_0) \tag{4.8}$$

A—readings of MA test instrument during measurement,

A_o*—initial reading of MA test instrument.*

In the case of a biaxial stress, it should be taken into account that the readings of the MA device are proportional to the differences of the main stresses:

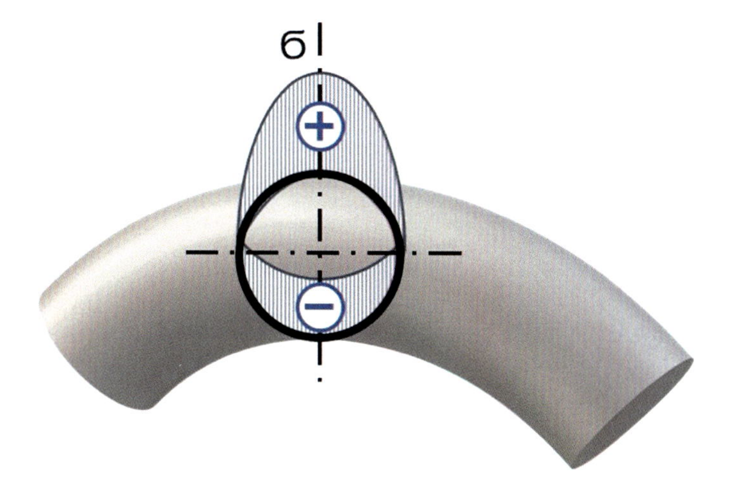

Fig. 4.28 Distribution of longitudinal stresses resulted from tensile load and bending in the cross-section of the main pipeline

$$\sigma_1 - \sigma_2 = T(A - A_0) \tag{4.9}$$

To determine the second component of the main stresses, another equation is required. For example, for main pipelines, such an equation is the Laplace equation for a thin-walled cylindrical shell:

$$\sigma_2 = pR/\delta \tag{4.10}$$

p *pressure in the pipeline,*
R *pipe radius,*
Δ *pipe wall thickness*

Determining the distribution of longitudinal stresses in the cross section of the main pipeline (Fig. 4.28) allows us to identify the causes of operational loads (for example, ground shift) and to monitor the dynamics of changes to prevent accidents.

4.8.4 Advantages and Limitations of MA Method

The advantages of the MA method are:

(1) Stress control.
(2) Simplicity and cheapness.
(3) High performance, the ability to perform measurement on painted structures.
(4) High informativeness.
(5) Mobility.

(6) Health, safety, and environmental friendliness.
(7) Ability to automate and create systems with many converters.

The limitations of applying the MA method are:

(1) Applicable only to ferromagnetic steels.
(2) Difficulties in controlling biaxial stress state and inability to control triaxial stress state.

4.9 Radiographic Testing

4.9.1 Method Fundamentals

In non-destructive testing, radiographic testing (RT) is used to determine:

(a) presence of discontinuities—cracks, cavities, solid inclusions, incomplete fusion, and lack of penetration (respectively, groups 1, 2, 3 and 4 of imperfections—Sect. 3.1), as well as root porosity (516—Sect. 3.1),
(b) imperfection coordinates,
(c) shape and size of imperfection,
(d) type of imperfection.

The method consists in radiographing of the welded joint by ionizing radiation and registration on the film (or screen) of zones of change in radiation intensity due to the presence of imperfections. In this case, electromagnetic waves in the frequency range above ultraviolet radiation, as well as flows of charged or neutral particles, are used as ionizing radiation.

Depending on the nature of the ionizing radiation the RT is divided into:

- X-ray—electromagnetic waves,
- gamma-ray—electromagnetic waves,
- beta—a stream of charged elementary particles—electrons (as well as positrons),
- neutron—a stream of uncharged elementary particles—neutrons.

The most practical applications for the testing of welded structures are X-ray and gamma-ray techniques. At the same time, X-ray technique has advantages in terms of control capabilities, including adjustment of radiation intensity, and ensuring high sensitivity.

The RT is based on physical effects, because of which ionizing radiation is generated, its intensity changes when passing through the control object (and in different ways in the defective and defect-free zone) and the residual ionizing radiation after passing through the object is recorded:

(1) **X-ray**—electromagnetic waves in the length range from 10 to 10^{-3} nm. On the scale of electromagnetic waves X-ray is located between ultraviolet radiation

and gamma-ray. X-ray occurs during the bombardment of a metal (anode) by a stream of accelerated electrons by two mechanisms:

- An accelerated electron is sharply slowed down (changes the speed and trajectory of motion) in the electrostatic fields of atomic nuclei and anode electrons. As a result of the energy loss of an accelerated electron, a *photon* (the smallest portion of an electromagnetic wave having high energy and having, therefore, the properties of a particle simultaneously with the properties of a wave) is emitted—**bremsstrahlung**. Some accelerated electrons are decelerated instantly on the surface of the anode, which corresponds to a photon with a minimum electromagnetic wavelength and maximum energy. Other accelerated electrons penetrate deep into the anode and gradually lose their energy. Thus, when electrons are decelerated, photons with very different energies will appear. Since the number of photons emitted per unit of time is very large, the spectrum (range of lengths of the components of electromagnetic waves) of bremsstrahlung is continuous.
- In addition to braking, an accelerated electron can knock an electron out of the internal electron shells of the anode atom. An empty space is formed on the inner electronic shell—a vacancy. Another electron from the outer shell of the anode atom (higher energy level) goes to the position of the vacancy (lower energy level). As a result of the transition of an electron from the outer shell of the atom to the inner, a photon is emitted—**characteristic radiation**. The photon energy is equal to the energy difference between the energy levels. Since each element of the Mendeleev's Periodic System of Elements has well-defined binding energies of electrons on the shells of an atom, each substance has a well-defined linear spectrum of characteristic radiation.

To obtain X-ray, an X-ray tube is used—a vacuum flask with a metal cathode (K) and anode (A) located inside (Fig. 4.29). A heating voltage U_h is supplied to the anode, which may be, for example, a tungsten filament. Free electrons in an X-ray tube are obtained around a heated cathode as a result of **thermionic emission**—the release of electrons from a solid (metal, carbides or borides) into free space (usually into vacuum or rarefied gas) when it is heated to a high temperature (above 900 K). The exit of electrons occurs as a result of an increase in energy upon heating to a level sufficient for the electron to perform the work function.

The resulting free electrons are accelerated by the difference in electric potential between the anode and cathode, which is created by a high accelerating voltage U_a, and they bombard the anode. X-ray emission occurs. The ability to control the accelerating voltage U_a in a wide range (from small units to hundreds of kV) provides the ability to control the intensity (stiffness) of X-ray. This is a significant advantage of the X-ray method compared to other types of Radiographic testing. Part of the energy released during electron braking is spent on heating the anode. Therefore, the anode is made of a metal with high thermal conductivity, for example, copper, and water cooling (C) is used. To form a beam of X-ray and gamma-ray (see below), a

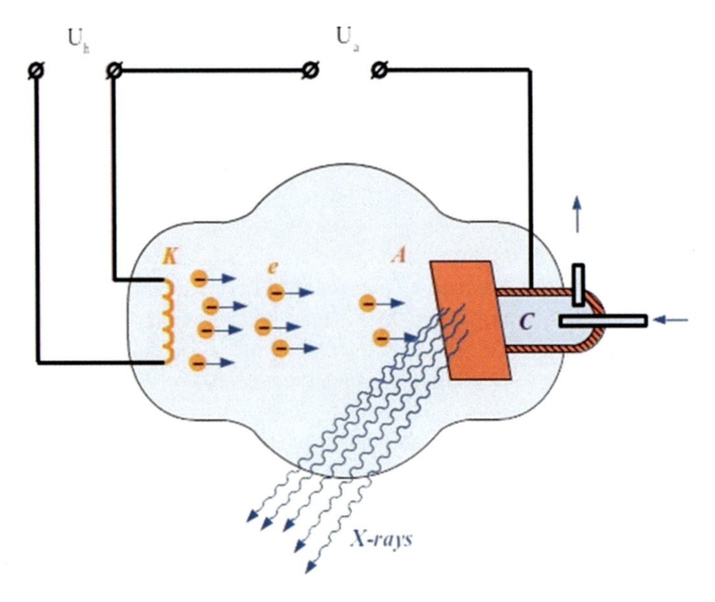

Fig. 4.29 X-ray tube structure: A—anode, K—cathode, e— free electrons, C—system of anode's cooling

collimator is used—an integrated or external device in the form of a lead cylinder (for example) with an aperture. The dimensions of the aperture determine the shape of the beam, the direction of the axis of the aperture—the orientation of the beam. The collimator absorbs scattered ionizing radiation to the maximum, which creates additional protection for personnel and prevents uncontrolled exposure of an X-ray film or other detector.

(2) **gamma-ray**—the shortest electromagnetic waves on the scale of electromagnetic waves in the range of length from 10^{-2} nm and below. By its nature, gamma radiation is similar to X-ray. gamma-ray arises from the radioactive decay of isotopes or because of the transfer of electrons from the outer high-energy shells of an atom. A photon is emitted, which has the highest energy on the electromagnetic wave scale. In non-destructive testing, isotopes are used as sources of gamma-ray:

- Thulium (170 Tm),
- Ytterbium (169a Yb),
- Selenium (75b Se),
- Iridium (192Ir),
- Cobalt (60Co).

A gamma defectoscope (Fig. 4.30) consists of a capsule with a gamma-ray source 1. The radioactive decay of isotopes in a gamma-ray source occurs continuously with the same intensity, therefore, when stored and transported (position 1a), it is in

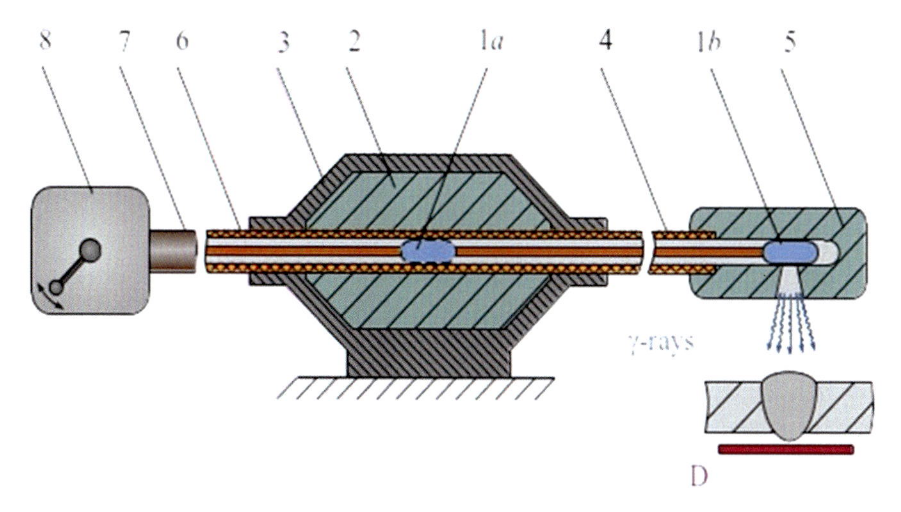

Fig. 4.30 The device gamma—defectoscope: 1—capsule with a source of gamma-ray (a—position during storage and transportation, b—position during testing), 2—container, 3—case, 4—capsule line, 5—external collimator, 6—cable, 7—a hose, 8—cable drive

a container 2 with thick walls, usually made of lead. The container is in the housing 3. During testing (position 1b), the capsule is moved through a flexible capsule line 4 to the collimator 5. The length of the capsule line can be more than 10 m. The capsule is moved using cable 6. The cable is driven inside the hose 7 by mechanical drive 8.

(3) **The high penetrating power** of X-ray and gamma-ray is due to:

- high energy of photon,
- short wavelength (commensurate, or less than interatomic distance in metals).

High penetrating ability determines the fundamental possibility of transmission of the test object in a wide range of thicknesses. Gamma-ray wavelength is shorter, and the photon energy is higher, therefore it is used for monitoring large thicknesses.

(4) **X-ray and gamma-ray attenuation** when passing through a metal is due to three main effects:

- Photoelectric absorption—the disappearance (absorption) of a photon in a collision with an atom and the transfer of energy to an electron. As a result, the electron leaves its shell and flies out of the atom with kinetic energy equal to the difference between the photon energy and the binding energy of the electron in the atom.
- Compton scattering—deviation of the trajectory (scattering) of a photon with its energy reduction when interacting with an electron. In this case, the decrease in the photon energy is transferred to the electron.

- Pair formation at high photon energy—the disappearance of a photon with the formation of a free electron e^- and positron e^+.

In the defect (discontinuity) zone, the thickness of the metal through which the X-ray and gamma-ray passes is smaller. Therefore, the radiation intensity in the defect zone is less decreased. The intensity of the residual radiation after passing through the test object in the defect zone is greater compared to the defect-free zone. This allows to identify the defect zone.

(5) **The impact on photosensitive materials (detectors)** of X-ray and gamma-ray is similar to the effect of light. The degree of impact on the detectors is proportional to the intensity of the residual radiation after passing through the control object. In the defect zone, the degree of exposure is greater compared to the defect-free zone. This allows to register the defect zone on the detector.

The following are used as detectors for X-ray and γ-control:

(a) X-ray film (F)—contains silver bromide, which is reduced to metallic silver (illuminated) under the influence of radiation and the subsequent development. Imperfections (discontinuities) appear in the form of blackouts, which have the shape and size of the imperfection—Fig. 4.31 (in some cases dimensions of imperfections can distort—see Sect. 4.9.4). Exception: inclusions with their density higher than that of the weld metal (for example, tungsten inclusions) which appear as bright spots.
(b) Storage phosphor imaging plate (IP)—a flexible polymer plate on which a photo stimulated luminescent material (phosphor) is applied. IPs are positioned like an x-ray film inside or outside the controlled product (Sect. 4.9.3). Exhibited on

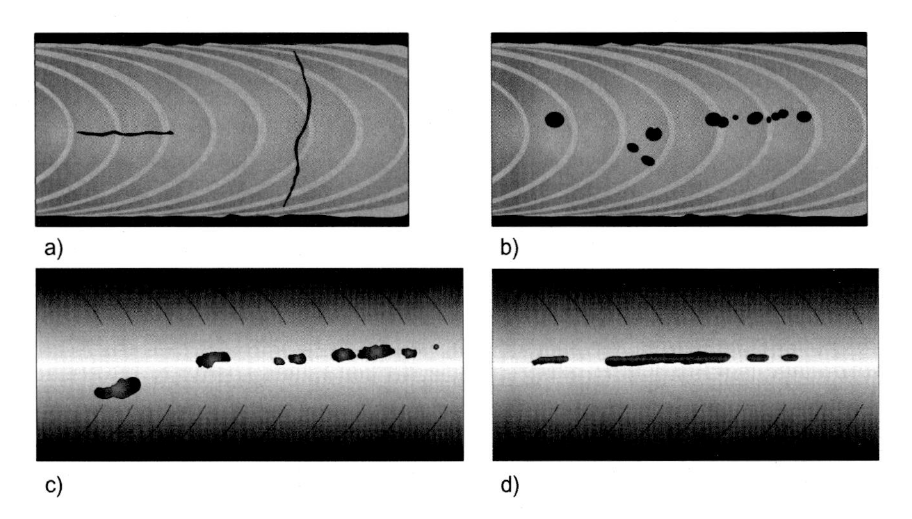

Fig. 4.31 Projections of the main types of defects on an x-ray film: **a** cracks (longitudinal and transverse), **b** pores (single, clusters and evenly distributed), **c** slag inclusions, **d** lack of fusion

conventional X-ray machines, but with reduced radiation energy and exposure time. Under the influence of X-ray or gamma-ray, the electrons of the phosphor crystals pass to higher energy shells of the atom and remain on them for a long time. Thus, a latent radiographic image is formed. IP is placed in a reader, which includes a laser scanner, an optical system (lens) and an analog-to-digital converter. When scanning under the influence of laser radiation, the electrons return to their original energy shells with the release of a photon of light (luminescence phenomenon). Glow of IP is proportional to the absorbed radiation. The resulting luminous image of the defect is transmitted through an optical system to an analog-to-digital converter and digitized. Unlike X-ray films, IP can be used in the light, they do not require a laborious development process. The most important difference between them is the possibility of reusing the IP (about 20 thousand exposures). IP image quality is superior to x-ray image quality. The IP dynamic range is 10 times higher than that of an X-ray film, which makes it possible to obtain images of products with a large difference in thicknesses on one image.

(c) Digital detector array or flat-panels (DDA)—an electronic screen (like the matrix of a digital camera) that converts x-ray or gamma-ray into an array of analog signals proportional to the radiation intensity. The analog signals are digitized and transmitted to the computer as a digital image. DDA have become popular with active areas up to 400 mm × 400 mm.

The development and improvement of IP and DDA created the conditions for the development of **computer radiography** (CR)—a method of RT which allows to obtain a digital 2D projection of defects through a single transmission of the test object. The CR system includes:

- X-ray source,
- IP (in combination with a reader) or DDA,
- processor, storage device, monitor.

Compared to X-ray film, computer radiography has the following advantages:

- obtaining a digital image of defects with the possibility of subsequent transmission, processing, presentation, and archiving using the entire variety of existing software,
- high detail and the ability to scale the image,
- the ability to receive images on a monitor screen in real time,
- significant reduction in exposure time, 50–60% reduction in radiation intensity,
- exclusion of expensive radiographic film,
- no need for chemical processing of images.

The direction of improving computer radiography is **computed tomography** (CT)—an RT that allows to get a digital 3D image of a defect by repeatedly scanning the object under inspection at different angles and subsequent calculations.

The resulting CT image of the imperfection can be presented in two formats: 3D-CT volume or a series of 2D-CT slices.

In the process of computed tomography, several projections of the image are taken from different angles. Recognition depends on the number of angles from which individual projections are obtained. Typically, the test object rotates 360° relative to the radiation source. The quality of computed tomography can be improved by increasing the number of projection scans.

CT system includes four main elements:

- X-ray apparatus,
- detector (DDA is most often used),
- devices for manipulating the control object, including any mechanical devices that affect image stability (it does not matter if the control object moves relative to the X-ray apparatus and detector or the X-ray apparatus and detector move relative to the control object),
- reconstruction/visualization system.

4.9.2 RT Area of Application

The RT is one of the most widely used non-destructive testing methods, despite the relatively high cost and biological hazard. If several methods are applied, the results of the RT are the main ones when making the final decision on the quality of the welded structure.

The RT is used for all types of steel, aluminum, copper, magnesium, titanium alloys.

Maximum thickness of tested steel structures:

- up to 150 mm—for x-ray inspection,
- up to 200 mm—for gamma control.

For non-ferrous metals and their alloys, the maximum thickness of the RT is greater.

4.9.3 X-ray and Gamma-Ray Techniques

The general steps of non-destructive testing are described in Sect. 4.10.

Details of the RT procedures are given in international standards:

- ISO 17636-1 [14]—X-ray and gamma-ray techniques with film,
- ISO 17636-2 [15]—X-ray and gamma-ray techniques with digital detectors,
- ISO 15708-2 [16], ISO 15708-3 [17]—Radiation methods for computer tomography.

The main features of the X-ray and gamma-ray techniques are the following.

Protection against ionizing radiation—national and international safety regulations must be strictly ensured.

Surface preparation—usually not required.

Radiogram identification—marking signs (for example, lead letters and numbers) are placed on each section of the object under control. Images of these signs should be displayed on a radiogram outside the control zone.

Marking on the testing object ensures the availability of reference points for accurate determination of the position of each radiogram (for example, reference, direction, identification, measuring belt).

When adjusting the exposure, the image quality indicator (IQI) is used—a sample with a simulator of imperfections of various sizes. Optimal exposure is determined by the minimum imperfection size that can be fixed. The most used IQIs are of three types:

- Wire (Fig. 32a)—a set of wires of various diameters fixed in a plastic cassette. Wire material: Fe—for control of steels and nickel alloys, Al—for control of aluminum alloys, Cu—for control of copper-based alloys. The diameter of the thinnest wire, the images of which can be seen on the film or screen, is taken as a measure of the achieved sensitivity (it should be borne in mind that the decision on the visibility of the wire is made by the operator, i.e. sensitivity measure is subjective).
- Lamellar (Fig. 32b)a plate of thickness T with three cylindrical holes with a diameter of T, 2 T, 4 T (T is 2% of the thickness of the test object). On a film or screen, a middle 2 T diameter hole should be clearly visible.
- Step-hole (Fig. 32c)—step plate with holes.

IQIs are made of metal or alloy, the basis of which in chemical composition is similar to that of a controlled weld.

IQI is placed on the surface of the test object from the side of the radiation source in the center of the control zone, on the base metal, next to the weld. When viewing IQI images on a radiogram, the number of the smallest distinguishable wire or hole is determined. An image of a wire is considered acceptable if its continuous section with a minimum length of 10 mm is clearly visible in a section with uniform optical density.

Radiographic testing is carried out in accordance with the schemes (Fig. 4.33), ensuring the closest possible location of the detector (F, IP, DDA—Sect. 4.9.1) to the control zone.

The axis of the radiation beam is oriented at the center of the control area, if possible, perpendicularly to the surface of the tested object. Minimization of the source-object distance is to be preferred.

The voltage on the tube (or the type of radiation source) and exposure time are the main parameters of the RT.

The voltage on the tube is set as low as possible to ensure high sensitivity control. In this case, in the picture the central part of the seam turns out to be white (or very light), and the base metal—black.

In computed tomography the additional control parameters are the number of exposures to obtain the projection, the gain during digitization and archiving number of integrations per projection, digitization gain, offset and binning.

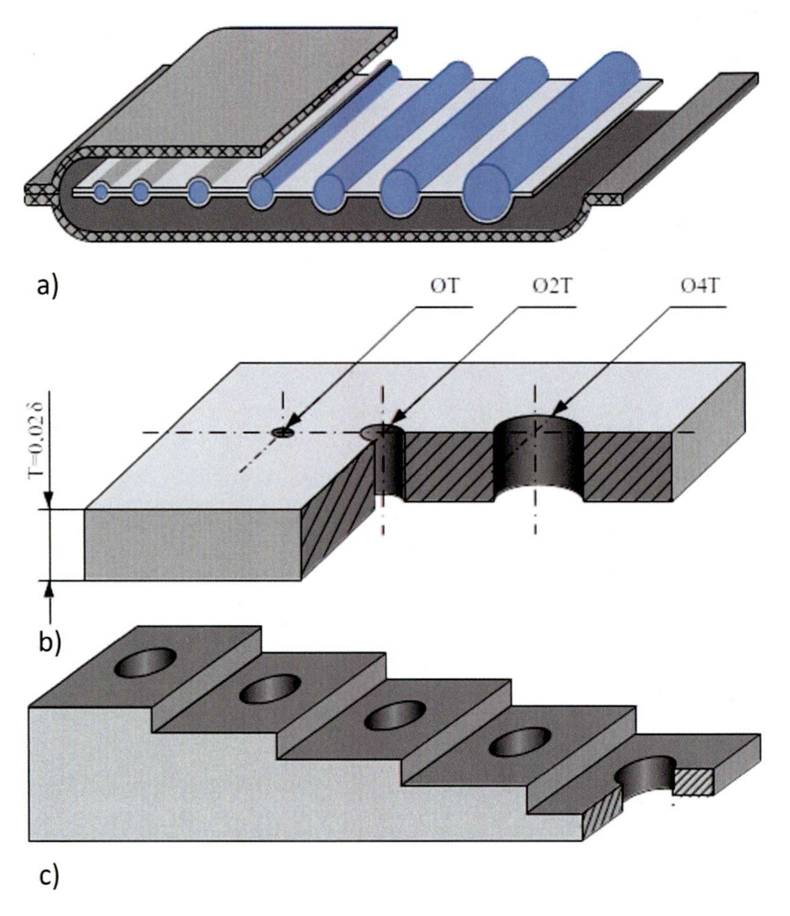

Fig. 4.32 Indicators of image quality: **a** wire, **b** plate (δ—thickness of the testing object), **c** step-hole

Minimum contrast-noise ratio (CNR) values should be achieved.

Interpretation of the results of the RT control with F or IP:

(a) The presence of an imperfection is determined by the presence of prints on the film (spots of various shapes and sizes or lines) black or dark (characteristic of discontinuities), and white or light (characteristic of tungsten inclusions).
(b) The imperfection coordinates are determined by the location of the imprint on the film relative to the marker.
(c) The shape and size of the imperfection is determined by the shape and size of the print on the film.
(d) The type of imperfection is determined by comparing the print on the film with the characteristic prints for different types of defects. An atlas of characteristic prints of types of imperfections is given, for example, in the IIW Radiographic reference.

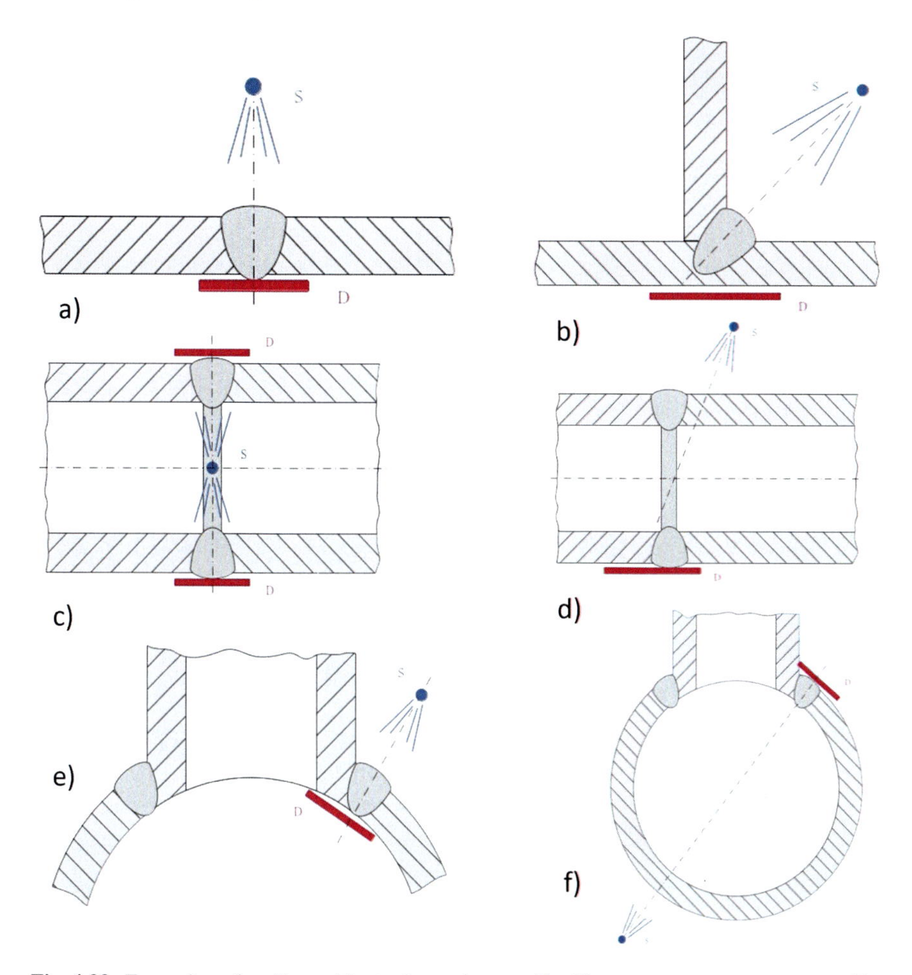

Fig. 4.33 Examples of radiographic testing schemes (S—X-ray or gamma-ray source, D—detector): **a** structures of small curvature, **b** fillet welds, **c** pipes of large diameter, transmission through one wall, **d** pipes of small diameter (up to 100 mm with a wall thickness of up to 8 mm), transmission on an ellipse, **e** flanges, transmission through one wall, **f** flanges, transmission through two walls

Reconstruction and visualization of the CT image from individual projections is the main step in computed tomography, which distinguishes this examination technique from other RT. During reconstruction and visualization of the image, the software that constantly improves the image is used, including different digital filter operations.

4.9.4 Advantages and Limitations of RT

The advantages of RT are:

(1) High detectability of defects.
(2) The ability to determine the type of imperfection.
(3) The ability to determine the shape and size of the imperfection.
(4) Objectivity and reliability of control. The presence of documented evidence of imperfections.
(5) Efficiency, the ability to control in real time.
(6) A wide range of controlled materials and thicknesses.

Limitations of the application of the RT are:

(1) Danger to health and the environment. The need for protective equipment. Disposal of X-ray films and reagents for their processing are environmental aspects of the organization.
(2) High cost and complexity of equipment for X-ray control. The need for special facilities.
(3) The need for access to the surface in the control zone.
(4) The limitation of computed tomography is the possibility of artifacts—artificial features which appear on the CT image but do not correspond to a physical feature of the sample.
(5) The accuracy of determining the size and location of an imperfection depends on the distance between the defect and the detector and the location of the radiation source relatively to the defect. This is due to an increase in the projection size of the defect on the detector with an increase in the distance between the defect and the detector, a shift in the projection of the defect on the detector when the radiation source is shifted relative to the defect, and a combination of these effects.

4.10 NDT Procedures. ISO 17635

General requirements for all NDT methods, procedures, and documented information are given in ISO 17635 [6]. In addition, standards related to testing method usually include details of control procedure specific to the method.

NDT procedures consist of the following stages.

(a) ***Agreement with interested parties***

In the case of NDT, interested parties are:

- organization that performs testing (process of testing),
- product manufacturer (processes of design and manufacturing),

- customer (if applicable).

Testing organization should be independent from the manufacturer (testing process should be independent from manufacturing) and its activities should be regulated by the quality management system (Sect. 1.5.8).

Items to be agreed between parties:

- aim and volume of control,
- level of control, acceptance level and criteria (Sect. 3.2),
- method of testing equipment set-up using standard samples (e.g., how the basic sensitivity level is set in US control device),
- stages of manufacturing, at which control should be performed, time and location of testing, control plan (if necessary),
- welded joints' parameters and heat-affected zone size data,
- requirements for accessibility, surface state and temperature,
- qualification of personnel,
- necessity of testing documented procedure, control, and final report requirements.

When defining time of testing one should take into consideration that control should be performed after all necessary stages of heat-treatment are finished. Welded joints in materials sensitive to hydrogen cracking (e.g., high-strength steels) should not be tested until minimally required time after welding (or time defined in the product specification) has passed.

Control of surface defects (e.g., VT) should be performed after the control of inner ones.

(b) ***Informing controllers***

Before the welded joint is tested controllers should receive the following information.

- Written testing procedure with information from item a) if necessary,
- basic material type and method of production (casting, rolling, forging, etc.),
- preparation of joint for welding and its dimensions,
- welding technology and any other relevant information about welding process,
- volume of control,
- time and volume of each post-welding heat-treatment,
- results of any previous basic material control (before or after welding),
- corrective actions in case unacceptable defects are detected,
- responsibility of subcontractor for coordination of control of parts manufactured by them.

(c) ***Test surface preparation***

The test surface should be plain. Undulation should not cause gaps between the sensor and the surface itself larger than 0.5 mm. Surface contamination should be removed (rust, loose sinter, welding spatters).

When control is performed by acoustic methods (US testing, acoustic emission testing) coupling medium is applied to the surface.

If surfaces are covered by coating, paint, deposited metal, etc. which cannot be removed a higher control level should be applied (e.g., d)—see Sect. 11 ISO 17640 [7]).

(d) ***Testing equipment set-up***

Sensitivity and range are set using standard samples with artificial defects.

If surface to be controlled is cylindrical (pipes, pressure vessels) as a rule the equipment should be profiled according to the test surface curvature.

(e) ***Conducting tests***

A starting point is chosen on the object's surface. Location of all defects is defined in relation to this point.

Scanning is performed according to the scheme defined by methodology or by standard(s) in the NDT method used.

Data analysis and interpretation is performed using the following steps.

- Testing data quality is defined—it is necessary to evaluate if discontinuities are depicted satisfactorily on the testing device monitor. For example, for UT PAUT method is used when data quality is evaluated by acoustic contact, time scan adjustment, sensitivity settings, rate of signal value to that of the noise, gap indication and completeness of data collected. The operator should decide if re-scanning is necessary. Such a decision demands high operator's qualification.
- Indicators are defined—signals corresponding to discontinuities.
- Classification of indicators—type of discontinuity is defined by signal amplitude, graph shape or other signal parameters.
- Coordinates and dimensions of discontinuity are defined.
- Discontinuities are evaluated according to acceptance criteria. If unacceptable defects are detected, corrective actions should be performed according to quality management procedure or to related standards used by the manufacturer (Sect. 1.5.13).

After removal of unacceptable defects, the second control of welded joint is performed. The second control uses the same requirements as the initial one.

(f) ***Test report and final report***

Reports are filled down by the personnel with specific permission and responsibility levels. Content of reports depends on standards and testing methods applied (Sect. 4.11).

4.11 Documented NDT Information

Documented NDT information includes the following:

(a) ***Documented control procedure***

The documented procedure describes stages and sequence of control operations. It ensures reproductivity of control results. It is developed according to requirements listed in the standard for particular control method(s) or according to product specification.

Documented control procedure includes the following information at minimum:

- aim and volume of control,
- testing methods,
- testing levels,
- qualification of personnel/requirements to training,
- requirements to equipment (including probes' characteristics, distance between probes, frequency of samples' choosing, etc.),
- standard samples for adjustment and calibration,
- equipment set-up,
- requirements to accessibility and surface state,
- base material control,
- control algorithm (or scanning scheme with description of probe location and movement, coverage, and volume of control),
- evaluation of indications,
- acceptance levels and/or registration levels,
- requirements for reports,
- problem of safety and environmental management.

Documented control procedure is approved and controlled according to quality management procedure (Sect. 1.5.11).

(b) ***Control plan***

A control plan is being developed in case when additional control is required, e.g., multiple NDT methods should be used, or control is performed repeatedly with one method. Plan defines location, sequence, and volume for each testing method.

(c) ***Testing protocol***

Protocol includes results of structure or part testing with one method. Information to be presented in the protocol:

- characteristics of test object (material and shape, geometry and dimensions, location of tested welded joint, sketch with dimensions if needed, details of welding and heat-treatment, surface state, temperature, and operating conditions),
- standard requirements for testing (technical conditions, norms, special requirements, etc.),

- place and date of testing,
- information about organization performing testing, testing operator and operator's qualification,
- manufacturer, model, and type of equipment (with serial numbers),
- manufacturer, type, characteristics of probe with serial number (if necessary),
- standard samples, including standard samples of the organization, with sketches, if necessary,
- expendable materials for testing (e.g., contact liquid, powder, etc.),
- control level and reference to the documented control procedure,
- volume of control,
- location of control zone (origin and coordinate system),
- scheme of testing (sketch),
- equipment set-up parameters,
- results of base metal control (in case it was performed),
- standards for acceptance levels,
- deviations from standards and contract requirements,
- coordinates of detected imperfections with details about related probe and control scheme,
- parameters of informative signal and information about defect type and size (if identification is possible),
- results of defects' assessment according to applied acceptance levels.

(d) ***Final report***

The final report summarizes results of control of batch of products or control of structure with multiple methods. It includes information listed in the control plan, at the minimum:

- test protocols (main part of final report),
- identification of parts and welded joints which were tested and/or reference to the document with IDs,
- documents describing coordinate system applied during control,
- references to particular test results, including statuses, such as not tested, accepted, not accepted,
- identification of personnel and organizations which took part in control,
- records about deviations from standard requirements to testing technology and acceptance levels.

References

1. CEN/TR 15135:2005 Welding—Design and non-destructive testing of welds
2. ISO 17635:2010 Non-destructive testing of welds—General rules for metallic materials
3. ISO 17637:2016 Non-destructive testing of welds—Visual testing of fusion-welded joints
4. ISO 9712:2021 Non-destructive testing—Qualification and certification of NDT personnel

5. ISO 5577:2017 Non-destructive testing—Ultrasonic testing—Vocabulary
6. ISO 17635:2016 Non-destructive testing of welds—General rules for metallic materials
7. ISO 17640:2018 Non-destructive testing of welds—Ultrasonic testing—Techniques, testing levels, and assessment
8. ISO 10863:2011 Non-destructive testing of welds—Ultrasonic testing—Use of time-of-flight diffraction technique (TOFD)
9. ISO 13588:2019 Non-destructive testing of welds—Ultrasonic testing—Use of automated phased array technology
10. ISO 22096:2007 Condition monitoring and diagnostics of machines—Acoustic emission
11. ISO 3452-1:2013 Non-destructive testing—Penetrant testing—Part 1: General principles
12. ISO 17638:2016 Non-destructive testing of welds—Magnetic particle testing
13. ISO 17643:2015 Non-destructive testing of welds—Eddy current testing of welds by complex-plane analysis
14. ISO 17636-1:2013 Non-destructive testing of welds—Radiographic testing—Part 1: X- and gamma-ray techniques with film
15. ISO 17636-2:2013 Non-destructive testing of welds—Radiographic testing—Part 2: X- and gamma-ray techniques with digital detectors
16. ISO 15708-2:2002 Non-destructive testing—Radiation methods—Computed tomography—Part 2: Examination practices
17. ISO 15708-3:2017 Non-destructive testing—Radiation methods for computed tomography—Part 3: Operation and interpretation

Printed in the United States
by Baker & Taylor Publisher Services